T0155862

Communications in Computer and Information Science 1592

More information about this series at https://link.springer.com/bookseries/7899

Jian Chen · Takashi Hashimoto · Xijin Tang ·
Jiangning Wu (Eds.)

Knowledge and Systems Sciences

21st International Symposium, KSS 2022
Beijing, China, June 11–12, 2022
Proceedings

Editors
Jian Chen
Tsinghua University
Beijing, China

Xijin Tang
CAS Academy of Mathematics and Systems Science
Beijing, China

Takashi Hashimoto
Japan Advanced Institute of Science and Technology
Nomi, Japan

Jiangning Wu
Dalian University of Technology
Dalian, China

ISSN 1865-0929 ISSN 1865-0937 (electronic)
Communications in Computer and Information Science
ISBN 978-981-19-3609-8 ISBN 978-981-19-3610-4 (eBook)
https://doi.org/10.1007/978-981-19-3610-4

This Springer imprint is published by the registered company Springer Nature Singapore Pte Ltd.
The registered company address is: 152 Beach Road, #21-01/04 Gateway East, Singapore 189721, Singapore

Preface

The annual International Symposium on Knowledge and Systems Sciences (KSS) aims to promote the exchange and interaction of knowledge across disciplines and borders to explore new territories and new frontiers. With more than 20 years of continuous endeavors, attempts to strictly define knowledge science may be still ambitious, but a very tolerant, broad-based and open-minded approach to the discipline can be taken. Knowledge science and systems science can complement and benefit each other methodologically.

The first International Symposium on Knowledge and Systems Sciences (KSS 2000), initiated and organized by Japan Advanced Institute of Science and Technology (JAIST), took place in September of 2000. Since then, with collective endeavours illustrated in KSS 2001 (Dalian), KSS 2002 (Shanghai), KSS 2003 (Guangzhou), KSS 2004 (JAIST), KSS 2005 (Vienna), KSS 2006 (Beijing), KSS 2007 (JAIST), KSS 2008 (Guangzhou), KSS 2009 (Hong Kong), KSS 2010 (Xi'an), KSS 2011 (Hull), KSS 2012 (JAIST), KSS 2013 (Ningbo), KSS 2014 (Sapporo), KSS 2015 (Xi'an), KSS 2016 (Kobe), KSS 2017 (Bangkok), KSS 2018 (Tokyo) and KSS 2019 (Da Nang), KSS has emerged as successful platform where many scientists and researchers from different countries and communities in knowledge and systems sciences have congregated.

During the past 20 years, people interested in knowledge and systems sciences have aggregated into a community, and an international academic society has existed for 19 years. After a two year hiatus due to COVID-19, the KSS event restarted with the theme of "Systems Thinking to the Impact of Pandemic and Knowledge Support", taking place in an online mode during June 11–12, 2022. To fit that theme, four distinguished scholars were invited to deliver the keynote speeches.

- Wei Huang (Southern University of Science and Technology, China), "Meta-synthesis decision methodology of complex system management in big data era".
- Van-Nam Huynh (JAIST, Japan), "Integrating machine learning and evidential reasoning for user profiling and recommendation".
- Ray Ison (IFSR and Open University, UK), "Beyond COVID-19: Reframing the global problematique with STiP (systems thinking in practice)".
- Dimitrios Kiritsis (ÉPFL, Switzerland), "Cognitive digital twins".

KSS 2022 received 52 submissions from authors studying and working in China, Finland, France, Germany, Japan, Spain, and Switzerland, and finally 17 submissions were selected for publication in the proceedings after a double-blind review process. The co-chairs of international Program Committee made the final decision for each submission based on the review reports from the referees, who came from Australia, China, Japan, New Zealand, Switzerland, Thailand, and the UK.

To make KSS 2022 happen, we received a lot of support and help from many people and organizations. We would like to express our sincere thanks to the authors for their remarkable contributions, all the Technical Program Committee members for their time and expertise in paper reviewing on an extremely tight schedule, and the proceedings publisher, Springer, for various professional help. This is the 5th edition of KSS proceedings published as a CCIS volume after successful collaboration with Springer during 2016–2019. We greatly appreciate our four distinguished scholars for accepting our invitation to present keynote speeches at the symposium. Last but not least, we are indebted to the organizing group for their hard work and the financial support from the China Association for Science and Technology.

We were happy to hear the thought-provoking and lively scientific exchanges in the fundamental fields of knowledge and systems sciences during the symposium, and we hope you enjoy the proceedings.

June 2022

Jian Chen
Takashi Hashimoto
Xijin Tang
Jiangning Wu

Organization

Organizer

International Society for Knowledge and Systems Sciences

Host

CAS Academy of Mathematics and Systems Science, China

Co-organizer

Systems Engineering Society of China

General Chairs

Jian Chen	Tsinghua University, China
Zengru Di	Beijing Normal University, China

Program Committee Co-chairs

Takashi Hashimoto	Japan Advanced Institute of Science and Technology, Japan
Xijin Tang	CAS Academy of Mathematics and Systems Science, China
Jiangning Wu	Dalian University of Technology, China

Technical Program Committee

Quan Bai	University of Tasmania, Australia
Zhigang Cao	Beijing Jiaotong University, China
Jindong Chen	Beijing Information Science and Technology University, China
Yong Chen	Shanghai Jiao Tong University, China
Yu-wang Chen	University of Manchester, UK
Beibei Gu	CAS Computer Network Information Center, China
Chonghui Guo	Dalian University of Technology, China
Rafik Hadfi	Kyoto University, Japan

Takashi Hashimoto	JAIST, Japan
Wenhua Hu	Wuhan University of Technology, China
Bin Jia	Beijing Jiaotong University, China
Jie Leng	CAS Academy of Mathematics and Systems Science, China
Ruijun Li	Auckland University of Technology, New Zealand
Weihua Li	Auckland University of Technology, New Zealand
Yongjian Li	Tianjin University, China
Zhenpeng Li	Taizhou University, China
Yan Lin	Dalian Maritime University, China
Bo Liu	CAS Academy of Mathematics and Systems Science, China
Jiamou Liu	University of Auckland, New Zealand
Yijun Liu	CAS Institute of Science and Development, China
Jinzhi Lu	EPFL, Switzerland
Jing Tian	Wuhan University of Technology, China
Mina Ryoke	University of Tsukuba, Japan
Jingli Shi	Auckland University of Technology, New Zealand
Bingzhen Sun	Xidian University, China
Xijin Tang	CAS Academy of Mathematics and Systems Science, China
Mengyan Wang	Auckland University of Technology, New Zealand
Jiang Wu	Wuhan University, China
Haoxiang Xia	Dalian University of Technology, China
Jianwen Xiang	Wuhan University of Technology, China
Jiangning Wu	Dalian University of Technology, China
Nuo Xu	Communication University of China, China
Zhihua Yan	CAS Academy of Mathematics and Systems Science, China
Thaweesak Yingthawornsuk	King Mongkut's University of Technology Thonburi, Thailand
Xiao Yu	Wuhan University of Technology, China
Yu Yu	Nanjing Audit University, China
Yi Zeng	CAS Institute of Automation, China
Shuang Zhang	Dalian University of Technology, China
Wen Zhang	Beijing University of Technology, China
Xiaochen Zheng	EPFL, Switzerland
Xiaoji Zhou	China Aerospace Academy of Systems Science and Engineering, China

Sponsor

China Association for Science and Technology

Abstracts of Keynotes

Meta-synthesis Decision Methodology of Complex System Management in Big Data Era: A New Complementary Panoramic Decision Framework

Wei Huang[1,2]

[1] Southern University of Science and Technology, Shenzhen
[2] Xi'an Jiaotong University, Xi'an, China
huangw7@sustech.edu.cn

Abstract. The importance of micro, small and medium-sized enterprises in the economy and people's livelihood can often be summarized as "6789"; that is, they contribute more than 60% of GDP, more than 70% of the national invention patents, 80% of urban job positions and 90% of new jobs. However, "the financing of these enterprises is a worldwide theoretical and practical problem of decision making." This is because these enterprises natural have some weaknesses, mainly manifested as development uncertainty, information asymmetry and scale diseconomies.

How to effectively solve the financing problems of micro, small and medium-sized enterprises lies in how to help and encourage financial institutions to make lending decisions to these enterprises, and lending decisions to these enterprises is a typical complex system management and decision-making problem (with the nine main characteristics of the complex system management and decision-making problem).

Targeting the study of this complex system decision-making problem, we propose a "meta-synthesis decision methodology of complex system management" to effectively improve the efficiency and success rate of loan decision making, and try to solve the worldwide theoretical and practical problem of decision making.

Keywords: Meta-synthesis decision methodology of complex system management · Micro, small and medium-sized enterprises · Financing · Complex system management and decision-making

Integrating Machine Learning and Evidential Reasoning for User Profiling and Recommendation

Van-Nam Huynh

Japan Advanced Institute of Science and Technology, Japan
huynh@jaist.ac.jp

Abstract. User profiles that represent users' preferences and interests play an important role in many applications of personalization. With the rapid growth of multiple social platforms, there is a critical need for efficient solutions to learn user profiles from the information shared by users on social platforms so as to improve the quality of personalized services in online environments. Developing an efficient solution to the problem of user profile learning is significantly challenging due to difficulty in handling data from multiple sources, in different formats and often associated with uncertainty. In this talk, we will introduce an integrated framework that combines advanced Machine Learning techniques with evidential reasoning based on Dempster-Shafer theory of evidence (DST) for user profiling and recommendation. Two instances of the proposed framework for user profile learning and one instance for multi-criteria collaborative filtering will be demonstrated with experimental results and analysis that show the effectiveness and practicality of the developed methods. Finally, some directions for future research will be highlighted.

Beyond COVID: Reframing the Global Problematique with STiP (Systems Thinking in Practice)

Ray Ison[1,2]

[1] President IFSR
[2] Professor of Systems, ASTiP Group, The Open University, UK
ray.ison@ifsr.org

Abstract. We live in unprecedented times, a period new to human history, the Anthropocene-world. From 2020 all have been subjected to the perturbations of pandemic, economic disruption, war and civil unrest and changes in whole-Earth dynamics associated with a human-induced Anthropocene. These perturbations can be understood as wave-fronts breaking on the shore of our historical ways of thinking and acting, buffered only by our historically derived institutions (norms, rules) and the governance systems we have invented. Collectively, we must take seriously the question: what purposeful action will aid human flourishing, create and sustain a viable space for humanity, in our ongoing co-evolution with the Anthropocene-Biosphere?

In earlier times others have articulated similar concerns about the human predicament. Turkish-American cybernetician, Hasan Özbekhan (1970)[1], introduced the idea of the 'global problematique' to refer to the 'bundle of problems' confronting humanity; this 'bundle' of systemically related problems persists. In fact it has become worse. Drawing on Özbekhan's report and arguments presented in detail in Ison & Straw (2020)[2], the case for innovation in ways of knowing realised by institutions, praxis and governance systems is posed. Practitioners of knowledge science and systems science have to consider reflexively the traditions of understanding out of which the think and act.

Members of the audience and the wider knowledge science and systems science community are invited to consider what actions can be taken within and by the cybersystemic 'community' to enable greater solidarity based on mutual appreciation and respect for differences-that-make-a-difference as a sought-after contribution to beneficial transformative action.

Keywords: Cybersystemics · Systems thinking in practice · Problematique · Systemic governance · Institutional innovation

---

[1] THE CLUB OF ROME. THE PREDICAMENT OF MANKIND. Quest for Structured Responses to Growing World-wide Complexities and Uncertainties. A PROPOSAL. 1970.

[2] Ison, R L & Straw, E. 2020. The Hidden Power of Systems Thinking. Governance in a climate emergency, Routledge, Abingdon.

Cognitive Digital Twins

Dimitrios Kyritsis

EPFL, Switzerland
dimitris.kiritsis@epfl.ch

Abstract. As a key enabling technology of Industry 4.0, Digital Twin (DT) has already been applied to various industrial domains covering different lifecycle phases of products and systems. To fully realize the Industry 4.0 vision, it is necessary to integrate multiple relevant DTs of a system according to a specific mission. This requires integrating all available data, information and knowledge related to the system across its entire lifecycle. It is a challenging task due to the high complexity of modern industrial systems. Semantic technologies such as ontology and knowledge graphs provide potential solutions by empowering DTs with augmented cognitive capabilities. The Cognitive Digital Twin (CDT) concept has been recently proposed which reveals a promising evolution of the current DT concept towards a more intelligent, comprehensive, and full lifecycle representation of complex systems. In this keynote speech the concept of CDT and its key features will be presented, together with some application cases developed in collaborative projects with EU industries.

Contents

Complex Systems Modeling and Knowledge Technologies

Data Mining and Machine Learning

PM2.5 Spatial-Temporal Long Series Forecasting Based on Deep Learning and EMD

Qiang Zhang(✉), Guangfei Yang(✉), and Erbiao Yuan

Institute of Systems Engineering, Dalian University of Technology, Dalian, China
18742519725@163.com, gfyang@dlut.edu.cn

Abstract. With the accelerated urbanization and industrialization, air pollution has become an important issue that affects the daily economy as well as development. By the formation of the monitoring network, how to forecast PM2.5 pollution more accurately has become a more important issue. In the field of PM2.5 pollution forecasting, a series of results have emerged so far. However, these methods do not perform well in long-term forecasting due to the spatial and temporal variability of PM2.5, and existing deep learning models consider only time-series variation, ignoring the spatial aspects of PM2.5 dispersion and transport. By considering the influence of spatial-temporal information on PM2.5 prediction, we propose a model GAT-EGRU to predict PM2.5 long time series based on Graph Attention Network (GAT), Gated Recurrent Unit (GRU) and coupled with Empirical Modal Decomposition algorithm (EMD), which incorporates the spatial correlation as well as long-term dependence of PM2.5. The experimental results show that the GAT-EGRU model has more obvious advantages in predicting PM2.5 concentrations, especially for long time steps. This proves that the GAT-EGRU model outperforms other models for PM2.5 forecasting. After that, we verify the effectiveness of each module by distillation experiments.

Keywords: Air pollution forecasting · Deep learning · Spatial-temporal prediction · Empirical modal decomposition

1 Introduction

With the rapid development of economy and the continuous progress of industrialization, air pollution has become an important problem affecting daily life and economic development. Due to the excessive emission of various pollutants, the quantity of various air pollutants including particulate matter (PM), sulfur dioxide (SO_2) and nitrogen dioxide (NO_2) has increased significantly in recent years. One of the most harmful to human body is PM2.5. PM2.5 refers to particulate matter in ambient air with an aerodynamic equivalent diameter less than or equal to 2.5 μm. It can be suspended in the air for a long time. The higher its concentration in the air, the more serious the air pollution is. Although PM2.5 is only a small component in the earth's atmosphere, but it has an important impact on air quality and visibility. Compared with coarser atmospheric particulate matter, PM2.5 has a small particle size, large area, strong activity, and is easily accompanied by toxic and harmful substances (e.g., heavy metals, microorganisms,

J. Chen et al. (Eds.): KSS 2022, CCIS 1592, pp. 3–19, 2022.
https://doi.org/10.1007/978-981-19-3610-4_1

etc.), and has a long residence time and long transport distance in the atmosphere, thus having a greater impact on human health and atmospheric environmental quality [1]. How to accurately predict the concentration of PM2.5 in the future moment is of great significance for controlling heavy pollution weather, government decision planning and preventing respiratory diseases [2].

In recent years, scholars have conducted extensive research on how to accurately predict PM2.5 concentrations at future moments. The existing forecasting methods are mainly numerical simulation-based methods, econometric statistics-based methods, and machine learning-based methods. The principle of the numerical simulation-based approach is to use computers to model the atmospheric dispersion and transport processes to make PM2.5 predictions through numerical simulations. Typical numerical simulation models include nested air quality prediction modeling system (NAQPMS) [3], regional multi-scale air quality model (CMAQ) [4], meteorological and chemical model coupling model (WRF Chem) [5], etc. The models of numerical simulations are able to give prediction results by changes in weather conditions. However, the numerical simulation model is highly dependent on data resources (real-time meteorological data and constantly updated emission inventory), long calculation time delay and difficult to capture the nonlinear relationship between variables, and thus have low accuracy. The method based on econometric statistics [6] predicts PM2.5 by finding the linear relationship with other pollutants and weather variables based on this relationship. However, this method requires artificial selection of variables and the study considers fewer influencing factors, which makes it difficult to capture the nonlinear relationship between variables and PM2.5, and the obtained prediction results are time-sensitive and can only be predicted in the short-term range. Machine learning based approaches are mainly divided into traditional shallow models as well as deep learning models. Deep learning neural networks have better predictive performance compared to traditional machine learning methods. The main reason is that the deep learning method can not only model more variables, but also capture the long-term and short-term characteristics of historical data. However, commonly used deep learning methods such as traditional recurrent neural networks (RNN) suffer from loss of temporal information [7] in PM2.5 long-term prediction, and convolutional neural networks [8] (CNN) tend to lose spatial information when applied to PM2.5 prediction, which all lead to their low accuracy in PM2.5 long-term prediction.

Based on the above research, we propose the GAT-EGRU model, which takes into account the spatial correlation as well as the long-term dependence of PM2.5. We built a graph data structure of cities, where nodes represent cities and edges represent interactions of PM2.5 between cities. The meteorological information of each city is used as a characteristic of the city nodes, and the association between cities is determined according to the likelihood of PM2.5 transmission between them. On the basis of the graph, the attention mechanism is used to determine the contribution of surrounding city nodes to the PM2.5 of the target city, so as to learn the spatial transmission information of PM2.5 between cities. Secondly, by performing empirical modal decomposition (EMD) of the urban PM2.5 time series and inputting the decomposed information into the gated recurrent unit (GRU), the spatial transmission information and sequence information are used to capture the spatial and temporal diffusion process of PM2.5. We perform up to 96 h of prediction on the experimental dataset and demonstrate that the proposed method has advantages in PM2.5 long spatio-temporal series prediction.

The article in this paper is organized as follows: the second part of this paper introduces our research area and data, discusses our basic research methods, and introduces the model GAT-EGRU in detail. The third part of the paper describes the experimental procedure as well as the experimental results, discusses the evaluation and comparison of the performance of different methods and models using metrics such as MAE and MSE, and the fourth part of the paper summarizes the contributions of our study and provides an outlook for further research in the future.

2 Methodology

In this section, we firstly describe the data used in this paper as well as the study area and define the prediction of PM2.5 using mathematical formulas. Secondly, based on the data, we constructed a graph data structure of PM2.5 based on the correlation of PM2.5, and based on this, we proposed a modeling method for PM2.5 long spatial-temporal series prediction.

2.1 Studied Area and Data

The study area chosen for this paper is 184 cities in the Beijing-Tianjin-Hebei region and the Yangtze River Delta region of China, which is the region with the highest prevalence of air pollution in China. Based on the previous work [9], we chose urban PM2.5 data from 2015–2018 as well as meteorological data, and recorded data every 3 h. Among them, PM2.5 data are obtained from national control sites, while meteorological data are from the European Centre for Weather Forecasting (ECMWF)[1] Climate Reanalysis Data ERA5[2] large dataset, which is inverse by satellite data and corrected by multi-source data reanalysis.

2.2 Problem Definition

PM2.5 prediction problem is a typical spatio-temporal series prediction problem [10]. Compared with time series prediction, spatio-temporal series prediction problem should not only consider the passage and connection of data time, but also consider the spatial correlation between nodes [11]. Next, we convert the PM2.5 prediction problem into a mathematical formulation. Suppose there are a total of N city nodes at time t. For city node i ϵ N, its PM2.5 concentration is x_i^t and its surrounding associated cities form a set M $= \{l_1, l_2, l_3, \ldots\ldots, l_m\}$, M ϵ N. For city node i ϵ N, there exists a node attribute matrix $P_i^t \in R^p$ at moment t, where p represents the number of meteorological attributes. Meanwhile, we decompose the time series of urban nodes by empirical modal decomposition (EMD) to obtain s subseries, then the values of each node's subseries at moment t are ${}_1u_i^t, \ldots\ldots, {}_su_i^t$, which constitute the EMD matrix U_i^t. Then, for the target node PM2.5 prediction problem can be defined as the following form:

[1] https://www.ecmwf.int/

[2] https://climate.copernicus.eu/climate-reanalysis.

$$\left[x_i^t, U_i^t, P_i^t, x_{l_1}^t, P_{l_1}^t, \ldots\ldots, x_{l_m}^t, P_{l_m}^t, x_i^{t+1}, U_i^{t+1}, P_i^{t+1}, \ldots\ldots x_{l_m}^{t+1}, P_{l_m}^{t+1}, \ldots\ldots, x_{l_m}^{t+\tau-1}, P_{l_m}^{t+\tau-1}\right]$$
$$\rightarrow f(\cdot) \left[x_{l_i}^{t+\tau}, x_{l_i}^{t+\tau+1}, \ldots\ldots, x_{l_l}^{t+T}\right].$$
$$\text{where } \tau \in [1, \ldots\ldots, T] \tag{1}$$

where $f(\cdot)$ is obtained by iterating our model for T steps:

$$f(\cdot) = \underbrace{g(\ldots g(g(\cdot)))}_{T times} \tag{2}$$

For training, this paper uses root mean square error as the loss function, assuming that the predicted value is $[\widehat{x_i^1}, \ldots, \widehat{x_i^T}]$ and the true observed value is $[x_i^1, \ldots, x_i^T]$, then:

$$MSEloss = \frac{1}{T}\sum_{t=1}^{T}\left(\widehat{x_i^T} - x_i^T\right)^2 \tag{3}$$

2.3 Graph Construction

Since PM2.5 receives multiple factors and this influence is often spatially correlated[12], in order to improve the accuracy of prediction, we will construct the graph based on the data as well as the migration and diffusion law of PM2.5 itself, using weather variables as attributes of nodes and edges, and then learn the migration and diffusion process of PM2.5 on the basis of the graph to get the spatial PM2.5 pollųtion we need correlation.

Node Attributes. The meteorological conditions of the city can affect the diffusion transmission of PM2.5 pollution. The meteorological attributes selected in this paper are shown in the Table 1, and these meteorological attributes can be used as characteristics of the nodes [13, 15].

Table 1. Meteorological Attributes of Nodes

Variables	Unit
Boundary layer height	m
K-index	K
Wind speed component u	m/s
Wind speed component v	m/s
Surface Temperature	K

(continued)

Table 1. *(continued)*

Variables	Unit
Relative Humidity	%
Total precipitation	m
Surface pressure	Pa

Adjacency Matrix. The distance between two cities and the height of the mountains between the cities can determine the degree of association of PM2.5 between the two cities [14], and we calculate the Adjacency Matrix based on the following formula:

$$A_{ij} = H\left(d_\theta - d_{ij}\right) \cdot H\left(m_\theta - m_{ij}\right), \text{ where}$$
$$d_{ij} = \rho_i - \rho_j \tag{4}$$
$$m_{ij} = \sup_{\lambda \in (0,1)} \left\{h\left(\lambda\rho_i + (1-\lambda)\rho_j\right) - max\left\{h(\rho_i), h\left(\rho_j\right)\right\}\right\}$$

where ρ_i represents the location (latitude and longitude) of node i, d_{ij} represents the distance between two nodes, and $H(\cdot)$ is the step function, $H(\mathrm{x}) = 1$ when and only when $\mathrm{x} > 0$. d_θ, m_θ are the thresholds of distance and height, here 300 km, 1200 m respectively. With the above constraints, PM2.5 can be transported and diffused only when the distance between two urban nodes is less than 300 km and when the mountain range between them is less than 1200 m [15].

By defining the node attributes and the adjacency matrix, we have successfully incorporated the meteorological conditions as well as the pollution dispersion pattern of PM2.5 into the graph, which will help us to make PM2.5 predictions later.

2.4 Model Structure

Our model consists of three main modules: the GAT module, which extracts the spatial information of PM2.5 on the basis of the graph constructed in 2.3; EMD module, which performs empirical modal decomposition of the original PM2.5 time series to obtain a number of subseries; GRU module, which integrates the spatial and temporal information obtained from the GAT module and the EMD module, performs PM2.5 prediction under the influence of weather conditions. Each of these three modules will be explained below (Fig. 1).

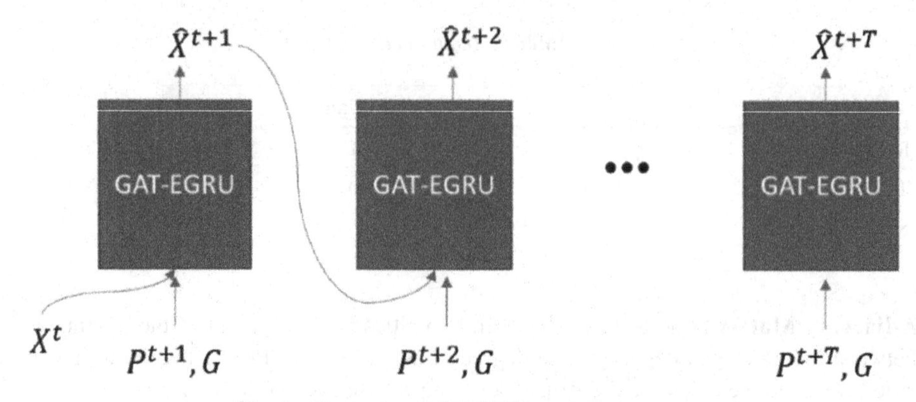

Fig. 1. Illustration GAT-EGRU model's process.

GAT Module

Due to the successful application of attention mechanism on various deep learning models, some scholars have started to combine the attention mechanism with graph neural networks. The graph attention network model was first proposed by velickovic et al. [16]. This model introduces an attention mechanism on the basis of graph neural networks, which greatly improves the prediction efficiency of the model. The principle of graph attention networks is to determine the weights of the different domains of the nodes in the process of model learning through the attention mechanism [17].

In the PM2.5 prediction problem, the graph attention network calculates the hidden state of each node in the graph through the attention mechanism, notices the neighboring nodes of the node, assigns different weight information to the neighboring nodes, and aggregates the feature information of the neighboring nodes according to the weight information. This approach enables efficient extraction of feature information of neighboring nodes and improves modeling flexibility, thus increasing the prediction efficiency of the model.

Firstly, we define the graph attention layer. At time node t, the input of the graph attention layer is the feature matrix of N nodes $P^t = \{P_1^t, \ldots, P_i^t, \ldots, P_N^t\}, P_i^t \in \mathrm{R}^p$, p is the number of node features, through the graph attention layer, we generate a new set of node feature matrix $P'^t = \{P'^t_1, \ldots, P'^t_i, \ldots, P'^t_N\}, P'^t_i \in \mathrm{R}^{p'}$. This means that for the input N nodes, each node has p' features, and through the graph attention layer, we obtain a higher dimensional node feature matrix P'^t.

After inputting the urban network graph structure of PM2.5 and training all the nodes, we will get the weight matrix $W \in \mathrm{R}^{p \times p'}$ of the graph, and the purpose of this matrix is to convert the low-dimensional features into high-dimensional features. Then we execute the self-attention mechanism a on each node: $a : \mathrm{R}^{p'} \times \mathrm{R}^{p'} \to \mathrm{R}$, and compute the attention coefficients between nodes and nodes by the self-attention mechanism:

$$e_{ij} = a\left(WP_i^t, WP_j^t\right) \tag{5}$$

This indicates the importance of node j to node i. Note that for node i, it is not necessary to calculate its attention coefficients with all other nodes, but only the attention coefficients of its neighboring nodes and it, to facilitate comparison, we normalize the attention coefficients using the SoftMax function:

$$\alpha_{ij} = softmax(e_{ij}) = \frac{\exp(e_{ij})}{\sum_{k \in M} \exp(e_{ik})} \tag{6}$$

where M is the set of node i's neighboring nodes.

The attention mechanism $a(\cdot)$ is a one-way feedforward neural network, the size of which is determined by the weight vector $\vec{a} \in R^{2p'}$, and is activated using the nonlinear activation function $LeakyReLU(\cdot)$, where the slope is taken as 0.2. We substitute to obtain the following equation:

$$\alpha_{ij} = \frac{\exp\left(\text{LeakyReLU}\left(\vec{a}^T\left[WP_i^t \| WP_j^t\right]\right)\right)}{\sum_{k \in M} \exp\left(\text{LeakyReLU}\left(\vec{a}^T\left[WP_i^t \| WP_k^t\right]\right)\right)} \tag{7}$$

where $\cdot^T$ denotes transposition and $\|$ denotes splicing operation.

With the above calculation, the output of each node can be obtained as:

$$P'^t_i = \sigma\left(\sum_{j \in M} \alpha_{ij} WP_i^t\right) \tag{8}$$

In order to make the prediction results more accurate and stable, we use a multi-headed attention mechanism [18], for K independent attention mechanisms for parallel operations, and then the results are stitched together, we can get the final spatial feature output results as:

$$P'^t_i = \|_{k=1}^K \sigma\left(\sum_{j \in M} \alpha_{ij}^k W^k P_i^t\right) \tag{9}$$

where α_{ij}^k denotes the result of attention coefficient normalization of the k th attention mechanism, then the final output feature matrix p'^t has Kp' features. With the introduction of the multi-headed attention mechanism, the attention learning of the graph neural network model will be more stable, focusing more on the important nodes and more important information, thus improving the prediction accuracy (Fig. 2).

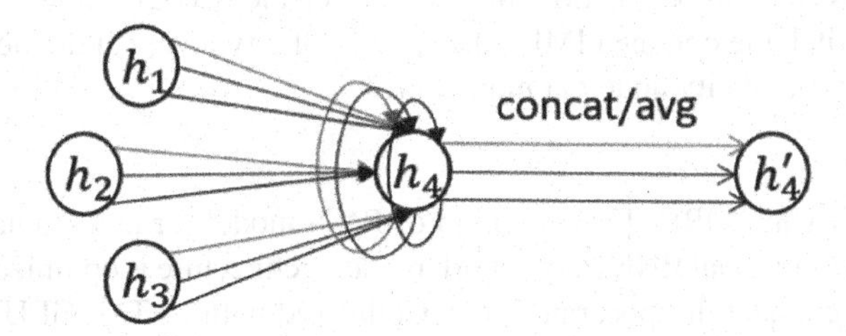

Fig. 2. An illustration of multi-head attention (with K = 3 heads) by node 1 on its neighborhood.

EMD Module

The main idea of empirical modal decomposition [19] (EMD), as a form of adaptive time series decomposition technique, is to use Hilbert-Huang transform [20] (HHT) to filter nonlinear and non-stationary time series data until a smooth series data is obtained. The key innovation of EMD is the introduction of the Intrinsic Mode Function (IMF). IMF can be well defined as a hidden oscillation pattern in a data sequence, since it can be non-smooth and can be amplitude or frequency modulated. IMF is defined as a function that satisfies the following two conditions:

- Throughout the data series, the number of extreme values (the sum of the number of maximum and minimum values) and the number of crossing zeros must be equal or differ by at most 1.
- At any point in the sequence, the average of the envelope defined by local maxima and the envelope defined by local minima is 0.

Based on the above two definitions, we decompose the PM2.5 time series $x_i(t)$ for a single node i by: determine all the extreme points of the sequence $x_i(t)$, fit the upper and lower extreme point envelopes (the upper and lower envelopes should include all data),and Find the average value of the upper and lower envelopes $m_i(t)$, subtract the average value from the original value to get the new sequence $h_i(t)$, and judge whether the sequence satisfies the definition of IMF function; determine whether the sequence satisfies the definition of the IMF function, and output the IMF if it does, and repeat the above operation with $h_i(t)$ until it satisfies the IMF definition, at which time $h_i(t)$ is the required IMF function $_k u_i(t)$;and Every time an order of IMF is obtained, it is subtracted from the original value until the last remaining part r_n is a monotonic sequence or a constant sequence.

After the decomposition by the above method, the original PM2.5 spatio-temporal sequence $x_i(t)$ of a single node is decomposed into a series of IMF functions and a linear superposition of r_n:

$$x_i(t) = \sum_{k=1}^{s} {}_k u_i(t) + r_n(t) \tag{10}$$

where s denotes the number of IMF functions after the decomposition is completed. After that, we stitch the obtained IMF subsequence of city node i at moment t to obtain the EMD matrix U_i^t of city node i at moment t.

GRU Module

Gated Recurrent Unit (GRU) [21] is a kind of RNN model for processing time series. Compared with traditional RNN, its network model architecture is optimized to alleviate the problem of gradient disappearance, etc. Compared with LSTM, GRU requires less computational resources and the effect is not worse than LSTM. GRU can selectively forget and retain the information of the past moment and the present moment through the gating unit, so as to realize the prediction of the future moment. The following is an explanation of the GRU module used in this paper.

At moment t, we obtain the spatial feature information p'^t_i that incorporates the surrounding nodes and the EMD matrix U^t_i. Then, the input of the spatio-temporal GRU component is obtained by stitching the spatial feature information p'^t_i as well as the EMD matrix U^t_i. In this way, the information we input into the GRU component contains both spatial feature information as well as temporal feature information, which will benefit us to improve the accuracy of the prediction because it takes into account not only the temporal diffusion but also the spatial transmission. We describe this operation and the ensuing GRU prediction by the following Eq. 11:

$$\begin{aligned} y^t_i &= \left[P'^t_i, U^t_i\right] \\ z^t_i &= \sigma\left(W_z \cdot \left[h^{t-1}_i, y^t_i\right]\right) \\ r^t_i &= \sigma\left(W_r \cdot \left[h^{t-1}_i, y^t_i\right]\right) \\ \tilde{h}^t_i &= \tanh\left(W \cdot \left[r^t_i \odot h^{t-1}_i, y^t_i\right]\right) \\ h^t_i &= \left(1 - z^t_i\right) \odot h^{t-1}_i + z^t_i \odot \tilde{h}^t_i \end{aligned} \tag{11}$$

where W_z, W_r, and W are all weight matrices that can be learned during the training process, y^t_i is the input, z^t_i is the update gate, r^t_i is the reset gate, and $\tilde{h}^t_i$ represents the spatio-temporal information at the current time, by updating the gate, the needed spatio-temporal information is kept and the unneeded ones are discarded to obtain the final spatio-temporal information h^t_i.

Finally, the final predicted values are obtained through the fully connected layer at:

$$\hat{X}^t_i = \Omega\left(h^t_i\right) \tag{12}$$

where Ω represents the fully connected layer.

3 Experiments

In this section, we will choose MAE and MSE indicators to verify the validity of the GAT-EGRU model by conducting experiments and comparing them with other prediction methods. Also, we will verify the effectiveness of each module through robustness experiments to determine the extent to which the model relies on spatio-temporal feature information as well as other weather variables.

3.1 Datasets

Based on the information described in Sect. 2.1, we formed a dataset of up to 4 years (2015/1/1 to 2018/12/31) including 184 cities and 8 weather variables selected based on previous studies (as in the Table 1) with a data granularity of 3 h to record one data point. In order to determine the correlation of these influencing factors associated with changes in PM2.5 concentrations, we first performed a correlation analysis on the data and calculated Pearson correlation coefficients between each weather variable and the PM2.5 series, and through the calculation we can know that there is a certain correlation between PM2.5 and each meteorological factor, and the calculation results are shown

Table 2. Pearson correlation coefficient of meteorological variables.

Variables	Pearson correlation coefficient
Boundary layer height	−0.34
K-index	−0.64
Wind speed	−0.53
Surface temperature	0.49
Relative humidity	0.15
Total precipitation	−0.44
Surface pressure	0.69

in the Table 2, where the PM2.5 series shows negative correlation with boundary layer height, k-index, wind speed and annual rainfall, and positive correlation with surface temperature relative humidity and surface pressure.

Subsequently, we preprocessed the dataset, including two parts. The first step is to normalize the data, which is very necessary in the algorithms of deep learning. The main reason for this is that the magnitudes of the individual weather variables are different, which can have a large impact on the convergence of the model and the accuracy of the final data. Here we use zero-mean normalization. After that, we divide the data after zero-meaning into a training set, a validation set and a test set in the ratio of 2:1:1 in order to check the predictive ability of the model under general remote settings.

3.2 Experimental Settings

Our task was to predict PM2.5 concentrations for the next 3, 12, 24, 48, 72, and 96 h based on the starting PM2.5 concentration and the weather variable data for the next 72 h, and to verify the validity of our proposed model, we selected the next models for comparison:

- nodesFC-GRU: This is a downscaled version of the proposed model GAT-EGRU. We achieve the predictive power of this model by replacing the first two modules of our model with a fully connected layer. With this model, we would like to know how much the spatio-temporal information we present from the first two modules actually brings an improvement to the predictive power of the model. This model can also be seen as an alternative to the CNN-like model.
- GAT: GAT means Graph Attention Network. GAT is used as a downscaled version of our model to determine the influence of GRU and EMD modules on this model. The version of GAT we use is based on the Pytorch-Geometric framework [22], which gives an advantage over other frameworks in terms of graph data processing and graph neural network operations. We only made changes to the last network layer to allow the model to work on this type of regression problem.
- GC-LSTM [23]: The model is a hybrid PM2.5 spatio-temporal prediction model based on graph convolutional neural network [24] (GCN) and long and short-term memory

[25] (LSTM), where the LSTM module does not differ much from our GRU module, so this model is designed to compare the advantages of GAT in extracting spatial information.

- HighAir [26]: The model is a hierarchical graph neural network-based air quality forecasting method. It considers the effect of spatial quality influencing factors on weather quality deterioration and constructs a graph data structure from a hierarchical perspective, using LSTM as a decoder. We chose this module to test the performance of the EMD module to ensure that is effective in GAT-EGRU.

Metrics. In order to better determine the performance of the evaluation model, two metrics, mean absolute error (MAE) and mean square error (MSE), are chosen in this paper as an evaluation of the model performance. MAE is the mean of the absolute value of the error between the observed and true values, and is used to describe the error of the predicted and true values [27]. MSE is a summed average of the squares of the difference between the observed and true values, and is generally used to detect the deviation between the predicted and true values of the model, which reflects the accuracy of the prediction [28]. The smaller these two metrics are, the better the predictive power of the model is and the more it meets our expectations.

Hyperparameter Optimization. In this paper, a Bayesian optimization [29] of the hyperparameters of the model is performed to obtain the optimal hyperparameter values of the model, and a fetch function is designed to avoid the local optimum; Bayesian optimization is based on a Gaussian process where the posterior is obtained by continuously adding new sampling points to update the prior distribution and taking into account the previous parameter information. Therefore, it is more reliable and efficient than random search and grid search. Our proposed method is optimized with the Adam optimizer with a batch size of 32 and a learning rate set to $1e^{-3}$, and an early stop is also used on the validation set.

Platform. All models were run on a single Tesla-V100-SXM2-32 GB GPU. We implement our model in Pytorch [30] and perform graphical calculations with Pytorch Geometric [22]. The data development platform is Dalian University of Technology's big data AIStation platform. The computational work in this paper was supported by the Research Center for Big Data and Intelligent Decision Making of Dalian University of Technology.

3.3 Results

We give the experimental results, and by careful analytical argumentation, the characteristics of the GAT-EGRU model aspects will be summarized in this paper. As the demand for prediction capability gradually increases, we have expanded the PM2.5 prediction step, but each model at each step is fully operational and processable on a single GRU. We will bold the best experimental results at each step size. The experimental results include the following four main components.

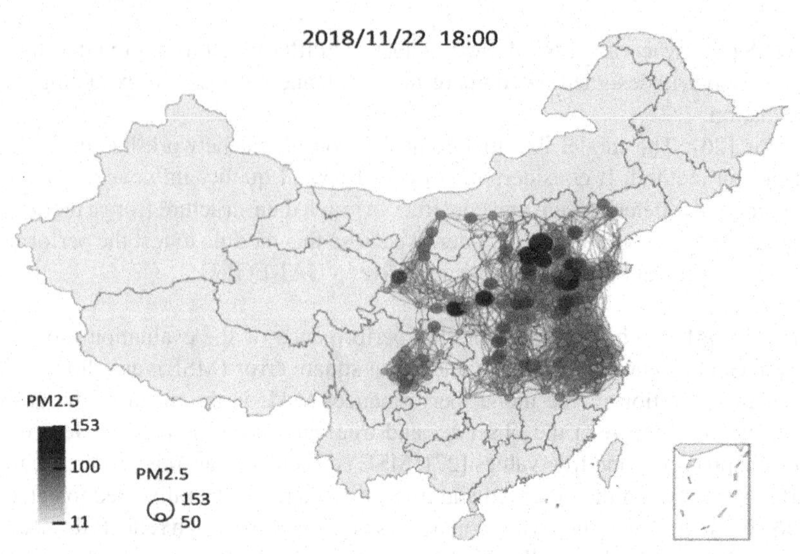

Fig. 3. Graph data structures. The circle size represents the PM2.5 concentration size and the edges represent the association between nodes.

Successful Construction of Graph Data Structures. First, we obtained the spatial association map of PM2.5 based on PM2.5 and the linkage rules, where the edges represent the association between city nodes, and the size of city nodes and the color shades represent the severity of pollution. We selected 18:00 on November 23, 2018, when pollution was high. From the Fig. 3, it can be seen that most of the pollution is concentrated in Beijing-Tianjin-Hebei and the Yangtze River Delta, and the PM2.5 in these regions has a strong correlation. Pollution in the southwest is less associated with other regions, and this pollution distribution is very reasonable considering its topography. Similarly, the diffusion of pollution in the northern regions is blocked by the influence of mountain ranges. The formation of contaminated areas illustrates the importance of correlation judgments, which is because we cannot judge the contamination status of other areas from areas that lack correlation. The difference in pollution between the north and the south [31] is shown by the presence of prominent pollution centers Polluted areas located in the east-central part of the country basically have centers of pollution which spread outward to form one or more large or small pollution areas. The formation of pollution areas in the south is very different, where the degree of pollution in various urban nodes is similar and there is no clear center of pollution, but the degree of pollution associated with this side is far greater than that in the north due to easy access. In addition to this, the pollution associations formed by the coastal cities are shown in the Fig. 3.

Model Performance and Evaluation. As shown in the Table 3, the number of GAT-EGRU wins was 11 out of 12 comparisons at 6 steps and for both metrics. This demonstrates that our model outperforms other models for PM2.5 prediction. As the prediction step size increases, other model errors begin to increase to higher levels. When the prediction step is 96, the MSE and MAE of GC-LSTM with HighAir are raised above 1 on

Table 3. Overall performance on all models. Best scores are in bold.

Model	GAT-EGRU	nodesFC-GRU	GAT	GC-LSTM	HighAir
Metric(h)	MSE MAE	MSE MAE	MSE MAE	MSE MAE	MSE MAE
3	**0.067 0.193**	0.114 0.205	0.157 0.199	0.194 0.218	0.225 0.348
12	**0.315** 0.442	0.353 **0.349**	0.370 0.397	0.445 0.517	0.578 0.696
24	**0.354 0.419**	0.401 0.433	0.428 0.454	0.543 0.589	0.688 0.726
48	**0.367 0.425**	0.474 0.452	0.496 0.517	0.895 0.648	0.794 0.819
72	**0.503 0.514**	0.591 0.539	0.578 0.594	0.984 0.801	1.092 1.114
96	**0.701 0.696**	0.812 0.744	0.766 0.751	1.088 0.991	1.605 1.828

the data processed with zero-mean processing. This means that at this step, both models essentially lose their predictive power, while the GAT-EGRU model remains at a lower level. When we put our eyes on the forecast step of 48 h, the time period that people often focus on in the real world, We find that the MSEs of the two models are 0.895 and 0.794, respectively, which are unacceptable values of this order of magnitude compared to our model of 0.367. This means that the deviation from the real value is too large to warn and guide people in their daily travel and life. The score between GAT-EGRU and nodesFC-GRU is 12:1, and numerically, these two models obtain similar scores on these two metric measures. However, we noticed that the metrics of nodesFC-GRU started to become larger and larger as the prediction step grew, for example, when the prediction step changed from 72 h to 96 h, its MSE and MAE increased by 0.221 and 0.205, while the increase of GAT-EGRU was only 0.198 and 0.182. This indicates that the spatio-temporal information extracted by our model plays their role in long-term prediction as the prediction step increases, and the model successfully captures the long-term dependence of the diffusion process. Similarly, the comparison with GAT also illustrates the excellent performance of EMD as well as GRU in capturing the time dependence, a capability that is necessary in the PM2.5 prediction process.

Comparison of Model Fine-Grained Analysis with Real Data. As a complement to the model performance and evaluation, we also compared the model to certain ground truth PM2.5 concentrations on the test set. The purpose of this comparison is to show why GAT-EGRU performs well numerically and in which areas it performs. We chose the target city, Dalian, as the object of comparison between our predicted and real values. This is due to its relatively even distribution of PM2.5 time as well as values, which can well reflect the performance of our model in predicting each value segment. As shown in the Fig. 4, it can be generally seen that the predicted trend of PM2.5 is consistent with the observed values, which validates the feasibility of our model to capture spatial and temporal variations. It can be seen that for extreme value points, our model tends to underestimate or overestimate the values, which is to some extent justified since our model does not have a module dedicated to outliers. Instead, our model is more accurate in predicting the trend for numerical intervals located in $[-1, 1]$. Overall, our model made a correct prediction for PM2.5 concentration values.

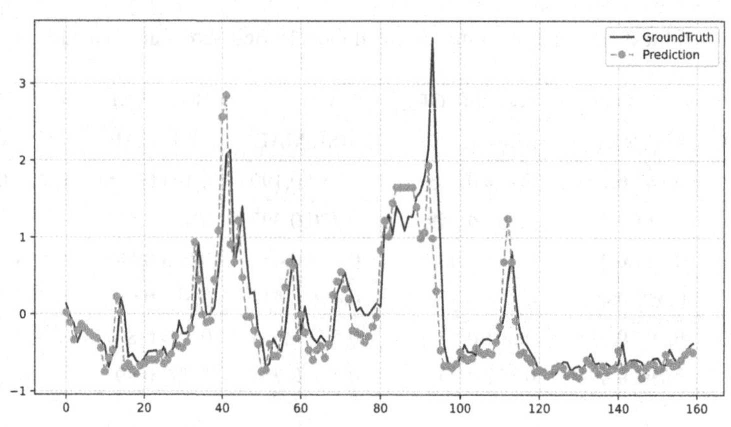

Fig. 4. Comparison of predicted and true values of GAT-EGRU model on the test set (node is Dalian city, prediction time is 24h, values are normalized)

Table 4. Robustness study for individual modules and weather variables

Model	GAT-EGRU	GAT-GRU	NodesFC-EGRU	GAT-EMD	GAT-EGRU-O
Metric(h)	MSE MAE	MSE MAE	MSE MAE	MSE MAE	MSE MAE
48	**0.367 0.425**	0.432 0.460	0.533 0.481	0.662 0.698	0.419 0.437
72	**0.503 0.514**	0.569 0.581	0.672 0.588	0.981 1.113	0.563 0.571
96	**0.701 0.696**	0.793 0.770	0.795 0.782	1.347 1.568	0.778 0.726

Analysis of Robustness Experimental Results. In order to verify the validity of the various modules of the model, we conducted distill ability experiments, which were designed to verify the robustness of the model as well as to determine the role of the various modules. The distillation experiments also allow us to verify the extent to which weather variables influence the performance of GAT-EGRU and thus determine the importance of this type of domain knowledge in terms of spatio-temporal long series prediction. The experimental results are shown in Table 4. Before proceeding to the analysis of the experimental results, the models in the robustness experiments are described below:

- GAT-GRU: We remove the EMD module from the model and use only the original PM2.5 sequence to stitch with the spatial information, this model is to verify the ability of EMD module in temporal information extraction;
- NodesFC-EGRU: This is different from NodesFC-GRU, where we keep the EMD module and now it is only the GAT module that is removed compared to the original model. This means that the spatial feature extraction part of the model is completely eliminated, leaving only the information fed to the GRU after processing by the EMD module, which is to verify that the spatial feature extraction is valid in our model;

- GAT-EMD: We replace the GRU in the GAT-EGRU with a fully connected layer. In fact, it is not just the elimination of a module, but the removal of our tool for capturing long-term spatio-temporal dependencies. We will use this model to determine the ability of GRU to capture long-term spatio-temporal dependencies, and thus determine that this module is highly desirable.
- GAT-EGRU-O: This is a judgment of domain knowledge, and we removed the most highly correlated weather variable, surface air pressure. We want to use this model to determine the importance of domain knowledge in such problems by judging that the selected weather variables play a role in the forecasting process, not just that previous studies judge them to be needed.

The Table 4 of experimental results shows that the performance of the model decreases significantly when any of the modules of the model are removed, which is consistent with our estimation. It is worth mentioning that when the GRU module is removed, the performance of the model is poor, which means that the model has long lost its predictive power at this point, and the time required to run the model is greatly increased from the experimental process. This means that simple variables as well as stacking of modules are of little help for nonlinear high-dimensional regression problems like PM2.5 long spatio-temporal series prediction. Throughout the iterative process, we must retain and remove information from the past sequence so that it is not disturbed by redundant information in the prediction of the future, which is the capability of the GRU model. Finally, the removal of variables brings a numerical decrease in the experimental results, but the performance of the model is still considered valid and this brings a reduction in computational complexity. The results of such an experiment lead to the following thoughts: How to choose the right weather variables to make trade-offs in terms of model performance and running speed may be an aspect we need to consider for real-world applications.

4 Conclusions

In this paper, we investigate an important real-world problem of how to accurately predict PM2.5 long spatio-temporal series and propose a method based on GAT-EGRU, a deep learning model coupled with an empirical modal decomposition algorithm. By observing the temporal and spatial characteristics of PM2.5, we constructed a spatial correlation map of PM2.5 and used GAT to capture the spatial correlation between different urban nodes to extract the spatial characteristics of cities; then the long sequence of PM2.5 is then decomposed and aggregated using an empirical modal decomposition algorithm to serve as a temporal signature of the PM2.5 transport process; finally, we stitch the spatio-temporal features obtained from the first two steps and input them into the integrated GRU to capture the spatio-temporal relationships in the PM2.5 pollution dispersion process and obtain the predicted values. We performed hyperparametric optimization of the model, trained it on a real dataset, and then validated it with sample data from the test set. Through extensive experiments, we demonstrate the success of the method on a real data set, proving the effectiveness of the method in improving the accuracy of PM2.5 long spatio-temporal series forecasting. Both in terms of numerical evaluation performance at all city nodes, comparison with ground truth concentrations, and results of robustness experiments, our model has a clear performance advantage in predicting PM2.5 long spatio-temporal series. Our study has the following limitations: firstly, the prediction performance for outliers is poor. This is because this paper does not include a module specifically for outlier prediction; second, the high spatio-temporal complexity of the model makes the prediction too slow when faced with large data sets, and we plan to optimize this in subsequent studies by introducing an outlier module and improving the complexity of the algorithm. In the future, we plan to investigate the interpretability of the model and its performance on other pollutants, and try to explore the characteristics of the PM2.5 spatio-temporal network.

Acknowledgments. This work was supported by the National Natural Science Foundation of China (42071273, 71671024, 71874021), Fundamental Research Funds for the Central Universities (DUT20JC38, DUT20RW301, DUT21YG119).

References

1. RenHe, Z., Li, Q., Zhang, R.: Meteorological conditions for the persistent severe fog and haze event over eastern China in January 2013. Sci. China Earth Sci. **57**(1), 26–35 (2013). https://doi.org/10.1007/s11430-013-4774-3
2. Zheng, Y., Liu, F., Hsieh, H.: U-air: when urban air quality inference meets big data. In: Proceedings of the 19th ACM SIGKDD International Conference on Knowledge Discovery and Data Mining, pp. 1436–1444 (2013)
3. Wang, Z., et al.: Modeling study of regional severe hazes over mid-eastern China in January 2013 and its implications on pollution prevention and control. Sci. China Earth Sci. **57**(1), 3–13 (2013). https://doi.org/10.1007/s11430-013-4793-0
4. Byun, D., Schere, K.L.: Review of the governing equations, computational algorithms, and other components of the Models-3 Community Multiscale Air Quality (CMAQ) modeling system (2006)
5. Grell, G.A., et al.: Fully coupled "online" chemistry within the WRF model. Atmos. Environ. **39**(37), 6957–6975 (2005)
6. Abhilash, M.S.K., Thakur, A., Gupta, D., Sreevidhya, B.: Time series analysis of air pollution in Bengaluru using ARIMA model. In: Perez, Gregorio Martinez, Tiwari, Shailesh, Trivedi, Munesh C., Mishra, Krishn K. (eds.) Ambient Communications and Computer Systems, pp. 413–426. Springer Singapore, Singapore (2018). https://doi.org/10.1007/978-981-10-7386-1_36
7. Hochreiter, S., Bengio, Y., Frasconi, P., Schmidhuber, J.: Gradient flow in recurrent nets: the difficulty of learning long-term dependencies. A field guide to dynamical recurrent neural networks. IEEE Press (2001)
8. LeCun, Y., Bottou, L., Bengio, Y., Haffner, P.: Gradient-based learning applied to document recognition. Proc. IEEE **86**(11), 2278–2324 (1998)
9. Yi, X., Zhang, J., Wang, Z., Li, T., Zheng, Y.: Deep distributed fusion network for air quality prediction. In: Proceedings of the 24th ACM SIGKDD International Conference on Knowledge Discovery & Data Mining, pp. 965–973 (2018)
10. Stanley, K.O., D'Ambrosio, D.B., Gauci, J.: A hypercube-based encoding for evolving large-scale neural networks. Artif. Life **15**(2), 185–212 (2009)
11. Pan, Z., Liang, Y., Zhang, J., Yi, X., Yu, Y., Zheng, Y.: Hyperst-net: hypernetworks for spatio-temporal forecasting. arXiv preprint arXiv:1809.10889 (2018)

12. Luan, T., Guo, X., Guo, L., Zhang, T.: Quantifying the relationship between PM 2.5 concentration, visibility and planetary boundary layer height for long-lasting haze and fog–haze mixed events in Beijing. Atmos. Chem. Phys. **18**(1), 203–225 (2018)
13. Li, X., et al.: Characteristics of particulate pollution (PM2.5 and PM10) and their spacescale-dependent relationships with meteorological elements in China. Sustainability-Basel **9**(12), 2330 (2017)
14. Wang, H., et al.: A multisource observation study of the severe prolonged regional haze episode over eastern China in January 2013. Atmos. Environ. **89**, 807–815 (2014)
15. Wang, S., Li, Y., Zhang, J., Meng, Q., Meng, L., Gao, F.: Pm2.5-gnn: a domain knowledge enhanced graph neural network for pm2.5 forecasting. In: Proceedings of the 28th International Conference on Advances in Geographic Information Systems, pp. 163–166 (2020)
16. Veličković, P., Cucurull, G., Casanova, A., Romero, A., Lio, P., Bengio, Y.: Graph attention networks. arXiv preprint arXiv:1710.10903 (2017)
17. Bahdanau, D., Cho, K., Bengio, Y.: Neural machine translation by jointly learning to align and translate. arXiv preprint arXiv:1409.0473 (2014)
18. Vaswani, A., et al.: Attention is all you need. Adv. Neural Inf. Process. Syst. **30** (2017)
19. Rilling, G., Flandrin, P., Goncalves, P.: On empirical mode decomposition and its algorithms. In: IEEE-EURASIP Workshop on Nonlinear Signal and Image Processing, 2003, pp. 8–11. Citeseer (2003)
20. Huang, N.E.: Hilbert-Huang Transform and its Applications, vol. 16. World Scientific. (2014)
21. Chung, J., Gulcehre, C., Cho, K., Bengio, Y.: Empirical evaluation of gated recurrent neural networks on sequence modeling. arXiv preprint arXiv:1412.3555 (2014)
22. Fey, M., Lenssen, J.E.: Fast graph representation learning with PyTorch Geometric. arXiv preprint arXiv:1903.02428 (2019)
23. Qi, Y., Li, Q., Karimian, H., Liu, D.: A hybrid model for spatiotemporal forecasting of PM2.5 based on graph convolutional neural network and long short-term memory. Sci. Total Environ. **664**, 1–10 (2019)
24. Kipf, T.N., Welling, M.: Semi-supervised classification with graph convolutional networks. arXiv preprint arXiv:1609.02907 (2016)
25. Hochreiter, S., Schmidhuber, J.: Long short-term memory. Neural Comput. **9**(8), 1735–1780 (1997)
26. Xu, J., Chen, L., Lv, M., Zhan, C., Chen, S., Chang, J.: Highair: a hierarchical graph neural network-based air quality forecasting method. arXiv preprint arXiv:2101.04264 (2021)
27. Chai, T., Draxler, R.R.: Root mean square error (RMSE) or mean absolute error (MAE)?–arguments against avoiding RMSE in the literature. Geosci. Model Dev. **7**(3), 1247–1250 (2014)
28. Allen, D.M.: Mean square error of prediction as a criterion for selecting variables. Technometrics **13**(3), 469–475 (1971)
29. Snoek, J., Larochelle, H., Adams, R.P.: Practical bayesian optimization of machine learning algorithms. Adv. Neural Inf. Process. Syst. **25** (2012)
30. Paszke, A., et al.: Pytorch: an imperative style, high-performance deep learning library. Adv. Neural Inf. Process. Syst. **32** (2019)
31. Wang, L., Liu, Z., Sun, Y., Ji, D., Wang, Y.: Long-range transport and regional sources of PM2.5 in Beijing based on long-term observations from 2005 to 2010. Atmos. Res. **157**, 37–48 (2015)

BugCat: A Novel Approach to Bug Number Categorization with Multi-modal Time Series Learning

Wen Zhang[1](✉), Rui Li[1](✉), Jiangpeng Zhao[1], Rui Peng[1], Yongwu Li[1], and Jindong Chen[2]

[1] College of Economics and Management, Beijing University of Technology, Beijing 100124, People's Republic of China
{zhangwen,pengrui19988,liyw}@bjut.edu.cn, {rui_li, zjp}@emails.bjut.edu.cn

[2] School of Economics and Management, Beijing Information Science & Technology University, Beijing, China
j.chen@amss.ac.cn

Abstract. It is of great importance to categorize the number of bugs over time for both software project managers and its end users. This paper proposes a novel approach called BugCat (i.e., Bug number Categorization) to categorize the bug numbers of software with multi-modal time series learning. The time series derived from the five modalities are used as the inputs of the proposed BugCat approach as the bug number, the day of the week, the day off, bug severity and bug priority. Then, the LSTM (Long Short-Term Memory) embedding is conducted on the five modalities of times series separately and, the concatenated vectors derived from data fusion on the five LSTM embeddings are used as the input of the full-connected neural network with ReLU (Rectified Linear Unit) activation to categorize the bug numbers of software. The extensive experiments with the Mozilla Firefox bug data demonstrate the superiority of the proposed BugCat approach over state-of-the-art techniques including multi-layer perceptron (MLP), fully convolutional network (FCN) and LSTM.

Keywords: Bug number categorization · BugCat · Multi-modal learning · LSTM · Software bugs

1 Introduction

Accurate bug number categorization has many benefits for both software project managers and software end users [1]. For software project managers, accurate bug number categorization has three benefits. The first is software quality, which can be informed by bug number categorization. It is generally accepted that the bug number categorization is inversely proportional to software quality. The second is resource allocation of software projects. Accurate bug number categorization will help project managers track software quality, and then help them to allocate and schedule limited testing resources. The third

J. Chen et al. (Eds.): KSS 2022, CCIS 1592, pp. 20–33, 2022.
https://doi.org/10.1007/978-981-19-3610-4_2

is that it can save the cost of software development and maintenance if we know bug number categorization in advance. For software end users, accurate bug number categorization will assist them to evaluate the quality of the software system objectively to decide when or whether to install the newly released version and, to obtain insights on the nature and development of software projects.

The bug number categorization mainly includes three reasons. The first reason is the delay of bug report submission and repair [2]. On the one hand, each bug usually has a scheduled expiration date. Project manager should ensure that as many problems as possible are completed within their respective deadlines to avoid adverse impact on the overall progress of the project. However, due to the high difficulty of bug repair, some bugs are delayed to be solved, which may lead to more bugs. On the other hand, Software developer may not submit a bug report immediately after discovering it, but defer the bug submission to the next working day. The second reason is repeated submission of bugs [3]. Although repeated submission of bugs will help developers to fix bugs more quickly and efficiently, the same bugs may exist in different periods of time, which will lead to waste of software testing resources and increase of costs. The third reason is the influence of noise [4]. The lack of high-quality data makes it difficult for the existing software defect prediction model to predict the accurate number of bugs in software modules under the influence of noise.

This paper proposes a novel approach called BugCat (i.e., Bug number Categorization) by considering five modalities (i.e., the bug number, the day of the week, the day off, bug severity and bug priority) to categorize the number of bugs using multi-modal time series learning. Traditional bug number categorization usually considers only the bug number series itself. However, our proposed BugCat approach takes the other four time series modalities into account besides the number of bugs. First, the LSTM embedding is conducted on the five modalities of times series separately. Second, the concatenated vectors derived from late data fusion on the five LSTM embeddings are used as the input of the full-connected neural network with ReLU activation to learn the complex relationships between five modalities. Finally, the last fully-connected layer with SoftMax activation is used to categorize the bug numbers of software. We tune the parameters to investigate the performance of BugCat for bug number categorization. We also devise ablation study to explore the effects of different modalities on BugCat approach. The extensive experiments with the Mozilla Firefox bug data show that the proposed BugCat approach outperforms state-of-the-art techniques on bug number categorization.

The rest of this paper is organized as follows. Section 2 presents related work. Section 3 illustrates the details of the proposed BugCat approach. Section 4 introduces experiment setting and results. Section 5 concludes the paper and indicates future work.

2 Related Works

2.1 Bug Number Prediction Based on Time Series

Traditional prediction methods based on time series mainly focus on modeling the number of historical bugs but ignore the other bug information such as bug priority and bug severity. For example, Wu et al. [5] applied X12 enhanced ARIMA to predict the bug number of Mozilla Firefox project with a sampling method of one month. Pati and

Shukla [6] applied Autoregressive Neural Network combing nonlinear autoregressive model (NAR) with embedded delay and feedback loop and Artificial Neural Network (ANN) to predict the bug number of Debian on time series. Pati et al. [7] compared Autoregressive Integrated Moving Average (ARIMA), back propagation neural network (BP-NN), and multi-objective genetic algorithm-based neural network (MOGA-NN) on 31 versions of ArgoUML to predict the software clone numbers on time series. Zhang et al. [8] proposed a SamEn-SVR approach combining sample entropy and support vector regression (SVR) to calculate the length and jump parameter of template vector through sample entropy, and then input template vector into SVR classifier to predict of bug number. This approach achieved optimal performance on four of the six versions of Mozilla Firefox data sets compared with ARIMA, X12 enhanced ARIMA, polynomial regression and artificial neural network (ANN).

2.2 Bug Number Prediction Based on Software Metrics

In addition to the prediction of bug number based on time series, many researchers use software metrics to predict the number of bugs. For instance, Andreou et al. [9] proposed stochastic Belief Poisson Regression Network method (SBPRN for short) to calculate the number of bugs. Santosh et al. [10] proposed to use integrated learning technology to predict the number of bugs in software modules of the current version based on the data of previous versions. Lei Qiao et al. [11] proposed deep learning-based software defect prediction approach and used 22 metrics to prediction the defects number of different modules. Xiao Yu et al. [12] found that regression algorithm is difficult to accurately predict the number of defects and suggested software testers use regression algorithms to rank modules for testing resource allocation, rather than predict the precise number of defects.

3 The Proposed Approach-BugCat

3.1 Overall Architecture

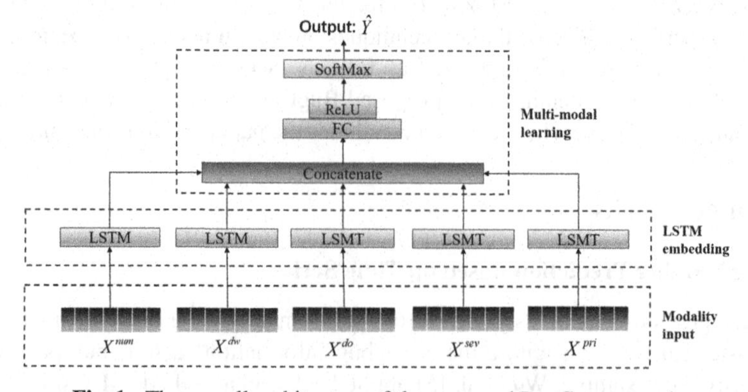

Fig. 1. The overall architecture of the proposed BugCat approach

The proposed BugCat approach consists of three sections: Modality input, LSTM embedding and Multi-modal learning. Figure 1 shows the overall architecture of the proposed BugCat approach. The inputs of this approach are the time series derived from the five modalities as the bug number (X^{num}), the day of the week (X^{dw}), the day off (X^{do}), bug severity (X^{sev}) and bug priority (X^{pri}). The output of this approach is bug number categorization ($\hat{Y}$) in the next time slots.

3.2 The Modality Input

Deep networks on multiple modalities have been successfully applied to supervised and unsupervised learning. The bug tracking system also contains a wealth of modalities information such as the bug number, the bug severity, and the bug priority etc. We introduce the five modalities used by the proposed BugCat approach as follows.

Modality 1: The bug number (X^{num}). To categorize the number of bugs, it is necessary to consider the bug number as the most important modality into the BugCat.

Modality 2: The day of week (X^{dw}). In our previous study [8], we found that the time series trend of bug number has periodicity for a week.

Modality 3: The day off (X^{do}). It is obvious that software developers and end-users submit more bugs on working days and fewer bugs on days off [8].

Modality 4: Bug severity (X^{sev}). The bug severity in Eclipse and Mozilla projects is divided into 7 levels such as Blocker, Critical and Major etc. [13].

Modality 5: Bug priority (X^{pri}). The bug priority in Eclipse and Mozilla projects is divided into 5 levels from P1 to P5 [14].

Given the historical bug modality information in the previous m time slots $x = [x_{t-m+1}, x_{t-m+2}, ..., x_t] \in \mathbb{R}^{N \times m}$, the bug number categorization problem can be formulated as learning a function $F(\bullet)$ for bug number categorization in the next time slot $\hat{y} = y_{t+1} \in \mathbb{R}^{N \times 1}$. For each modality time series input $X = \{X_1, X_2, ..., X_T\}$ with time interval τ, we use a time window of length m to reconstruct time series X into the shape of (samples, time steps), where the size of samples equal to $(T - m + 1)$, the size of time steps equal to m. For the proposed BugCat approach, there are five modality inputs as the bug number (X^{num}), the day of the week (X^{dw}), the day off (X^{do}), bug severity (X^{sev}) and bug priority (X^{pri}). Meanwhile, the constant time interval τ is one day. That is, we move forward the categorization with the step as one day.

3.3 The LSTM Embedding

The bug number categorization in future is correlated with the time dependence of each modality, and this correlation varies for each day. To capture this time-dependence, we devise a LSTM embedding with five LSTM layers. Figure 2 shows the architecture of the LSTM embedding. The inputs of LSTM embedding include the bug number (X^{num}), the day of the week (X^{dw}), the day off (X^{do}), bug severity (X^{sev}) and bug priority (X^{pri})

in accordance with Modality input section. The outputs of LSTM embedding include h_m^{num}, h_m^{dw}, h_m^{do}, h_m^{sev} and h_m^{pri}.

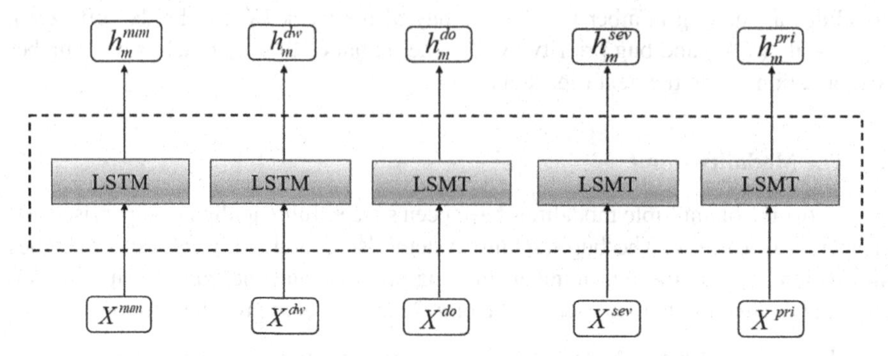

Fig. 2. The architecture of the LSTM embedding

3.4 Multi-modal Learning

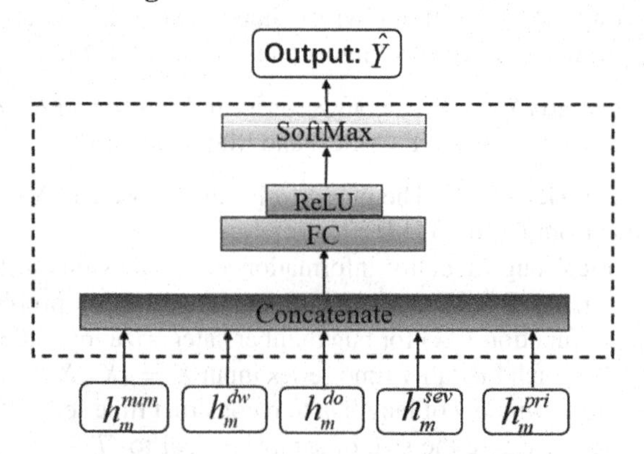

Fig. 3. The architecture of the Multi-modal learning

The bug number categorization is also correlated with complementarity between modalities, and this complementarity can be derived from the data fusion. To obtain the complementarity between modalities and perform multi-modal time series learning, we adopt multi-modal learning on a concatenation layer and two fully-connected layers. Figure 3 shows the architecture of the Multi-modal learning. The inputs of multi-modal learning are h_m^{num}, h_m^{dw}, h_m^{do}, h_m^{sev} and h_m^{pri}. The output of multi-modal learning is bug number categorization $\hat{Y}$. To obtain the complementarity between modalities, a concatenation layer is used to fuse five modalities and the inputs $\{h_m^{num}, h_m^{dw}, h_m^{do}, h_m^{sev}, h_m^{pri}\}$ are concatenated as fusion vector $v_k = [h_m^{num}, h_m^{dw}, h_m^{do}, h_m^{sev}, h_m^{pri}]$ at this layer.

Then, the vector v_k derived from data fusion are used as the input of the full-connected layer with ReLU (Rectified Linear Unit) activation to learn the complex relationships between five modalities and the output of this layer is Y_k^D as described in Eq. (1), where W_D is the weight matrix and b_D is the bias of the Y_k^D.

$$Y_k^D = \max(0, W_D v_k + b_D) \tag{1}$$

Finally, the vector Y_k^D is fed into the last fully-connected layer with SoftMax activation to categorize the number of bugs and the output vector $\hat{Y}_k$ as described in Eq. (2) with weight matrix W_{out} and bias vector b_{out}.

$$\hat{Y}_k = SoftMax(W_{out} Y_k^D + b_{out}) \tag{2}$$

3.5 The Loss Function

The training procedure of the proposed BugCat approach is a gradient descent process. In other words, the loss function $Loss(X_k) = Loss(Y_k, \hat{Y}_k)$ of $X_k = (X_k^{num}, X_k^{dw}, X_k^{do}, X_k^{sev}, X_k^{pri})$,where Y_k is a one hot vector of length three (For example, if bug number categorization is low, $Y_k = (1, 0, 0)$) and $\hat{Y}_k$ is a vector of length three whose sum is 1. Adding the loss of each training sample we can obtain the loss on the entire training batch as $Loss^{sum} = \sum_{k=1}^{|X|} Loss(X_k)$. Therefore, the weight matrices and bias vectors are updated iteratively as $W_l := W_l + \alpha \frac{\partial LOSS^{sum}}{\partial W_l}$ and $b_l := b_l + \alpha \frac{\partial LOSS^{sum}}{\partial b_l}$,where α is the learning rate and l denotes different layers mentioned above. The chain derivation rule is applied to update of neural network modal parameters iteratively with the Adam optimization algorithm.

Specifically, the proposed BugCat approach adopts the categorical cross-entropy loss function to categorize the number of bugs. The categorical cross-entropy loss function calculates the loss of a test instance by computing the following sum as shown in Eq. (3).

$$LOSS = -\sum_{i=1}^{C} Y_i * \log \hat{Y}_i \tag{3}$$

Here,$\hat{Y}_i$ represents the probability of i-th bug number categorization,Y_i represents the test instance whether belongs to i-th bug number level (if so, Y_i equals to 1; else, Y_i equals to 0), and the C is the number of bug number categorizations.

4 Experiment

4.1 Dataset

In order to investigate the performance of BugCat approach, we carry out extensive experiments on Mozilla Firefox bug dataset released in Mining Software Repositories 2010 (MSR2010). The Mozilla Firefox project has developed for more than 19 years and its bug tracking system contains abundant valuable information. Therefore, the bug reports from 23 September 2002 to 18 April 2009 are used to construct the benchmarking

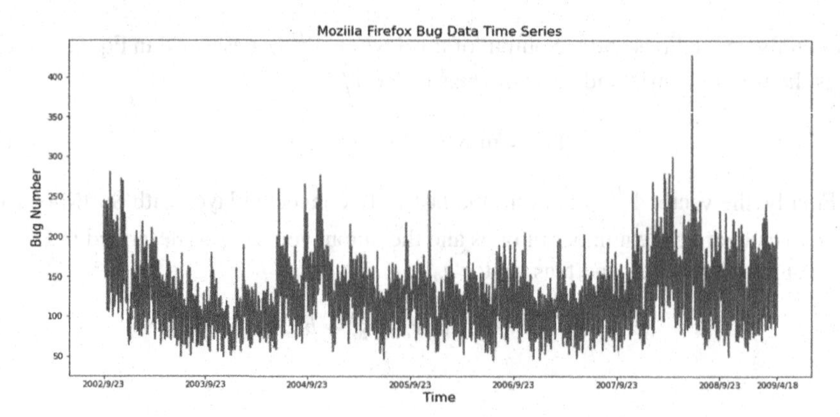

Fig. 4. The time series trend of the Mozilla Firefox bug number

data used in this research. Figure 4 shows the time series trend of the Mozilla Firefox bug number. First, we count the bugs by using the single day as the basic unit and collect 2400 days with 308002 unique bugs. The time series trend of Mozilla Firefox bug number is shown in Fig. 5. Second, in order to reduce the influence of software release updates on the experiment and to evaluate the performance of the proposed BugCat approach, we set 600 days as the interval to divide the whole benchmarking data into four subsets. Table 1 shows the bug number of each subset.

Table 1. The bug number of each subset

Subset	Time duration	# of bugs	# of days
1	23 September 2002–14 May 2004	73235	600
2	15 May 2004–4 January 2006	78512	600
3	5 January 2006–27 August 2007	67979	600
4	27 August 2007–18 April 2009	88276	600

Third, in order to make the categorizations of each subset be relatively balanced, we divide each subset into three categorizations at unequal interval and each subset consists of about 200 days. The split results are shown in Table 2.

Table 2. The bug number categorization distribution of each subset

Subset	Level	Interval	# of days	Ratio
1	Low	[0,100)	208	34.7%
	Middle	[100,135)	203	33.8%
	High	[135,)	189	31.5%
2	Low	[0,115)	203	33.8%
	Middle	[115,145)	198	33.0%
	High	[145,)	199	33.2%
3	Low	[0,100)	193	32.1%
	Middle	[100,130)	232	38.7%
	High	[130,)	175	29.2%
4	Low	[0,120)	187	31.2%
	Middle	[120,170)	222	37.0%
	High	[170,)	191	31.8%

4.2 Experiment Setting

For benchmarking datasets, we divide each subset into training set, validation set and test set with the proportion 8:1:1, i.e., we choose 80% of each subset as the training set to optimize the parameters in the proposed approach, 10% of each subset as the validation set to validate the model during the model training and 10% of each subset as the test set to evaluate the trained model's prediction performance. In addition, we partition each subset 10 times randomly and the performances of the experiments are averaged on the 10 repetitions.

4.3 Experimental Results and Discussion

This section shows the results of the BugCat approach compared with baseline methods on bug number categorization. We first propose three research questions to evaluate the performance of BugCat approach and then devise and conduct corresponding experiments to solve these questions.

RQ1: (Performance) How does the performance of BugCat compare to the baseline methods?
RQ2: (Parameter sensitivity) How to set parameter values of BugCat to achieve the best performance?
RQ3: (Ablation study) How do different modalities affect the bug number categorization?

4.3.1 Parameter Sensitivity Analysis

To investigate RQ1, we employ a greedy strategy to gradually tune the values of the three parameters. First, we tune the hidden units of LSTM embedding, with the other two parameters fixed at default values (i.e., 64 for hidden units of FC layer and 32 for batch size). Specifically, we tune the hidden units of LSTM embedding with values in the range of [4, 8, 16, 32, 64, 128, 256, 512], and train the model. Among all the values of hidden units of LSTM embedding, we select the optimal value whose corresponding model achieves the best performance. Second, we tune the hidden units of FC layer. Like the approach used to tune the first parameter, we conduct each value in the range of [4, 8, 16, 32, 64, 128, 256, 512], with the optimal hidden units obtained in the previous step and the default value of batch size. For all candidate hidden units of FC layer, we select the optimal value whose corresponding model achieves the best performance. Last, we tune the last parameter, batch size. The optimal values of the first two parameters are used in this step. Following the previous steps, we conduct each value in the range of [2, 4, 8, 16, 32, 64, 128, 256] and choose the value which makes BugCat achieve the best performance. According to the above three steps, three optimal parameters of the BugCat approach are obtained.

Table 3, 4 and 5 show the variation of BugCat's performance on accuracy under the different values for parameters. Obviously, BugCat achieves the best performance when the value of LSTM layers is 8, the value of FC layer is 256 and the value of batch size is 128.

Table 3. The hyper-parameter tuning of LSTM-units

Lstm-units	4	8	16	32	64	128	256	512
Accuracy	69.27%	**69.72%**	68.72%	67.70%	67.28%	66.31%	65.34%	64.73%

Table 4. The hyper-parameter tuning of FC-units

FC-units	4	8	16	32	64	128	256	512
Accuracy	68.17%	68.75%	68.83%	69.07%	69.25%	69.35%	**69.88%**	68.99%

Table 5. The hyper-parameter tuning of batch sizes

Batch_sizes	2	4	8	16	32	64	128	256
Accuracy	69.24%	69.35%	69.49%	69.54%	69.56%	69.67%	**70.06%**	69.51%

4.3.2 Baseline Comparison

In order to investigate RQ2, we select three comparison baselines, including LSTM (Long Short-Term Memory), multi-layer perceptron (MLP) and fully convolutional network (FCN) to compare their performances with BugCat. Specifically, compared with the BugCat approach which considers five modalities, these three traditional classifiers of bug number categorization only consider the bug number modality. Thus, in this question, we explore whether the BugCat approach that takes more modalities information into account performs better than baseline methods.

Figure 5 shows the results of BugCat compared with three baseline methods for five metrics on four subsets. It is obvious that the proposed BugCat approach achieves the best performance compared with other baseline methods. On accuracy measure, the BugCat approach, on average, has produced an improvement of 20.48% over the LSTM method, 9.21% over the MLP method and 7.93% over the FCN method. On macro precision measure, the BugCat approach, on average, has produced an improvement of 22.82% over the LSTM method, 10.69% over the MLP method and 9.31% over the FCN method. On macro recall measure, the BugCat approach, on average, has produced an improvement of 19.55% over the LSTM method, 9.19% over the MLP method and 7.43% over the FCN method. On macro F-score measure, the BugCat approach, on average, has produced an improvement of 22.64% over the LSTM method, 9.84% over the MLP method and 8.35% over the FCN method. On kappa measure, the BugCat approach, on average, has produced an improvement of 46.39% over the LSTM method, 18.76% over the MLP method and 15.98% over the FCN method.

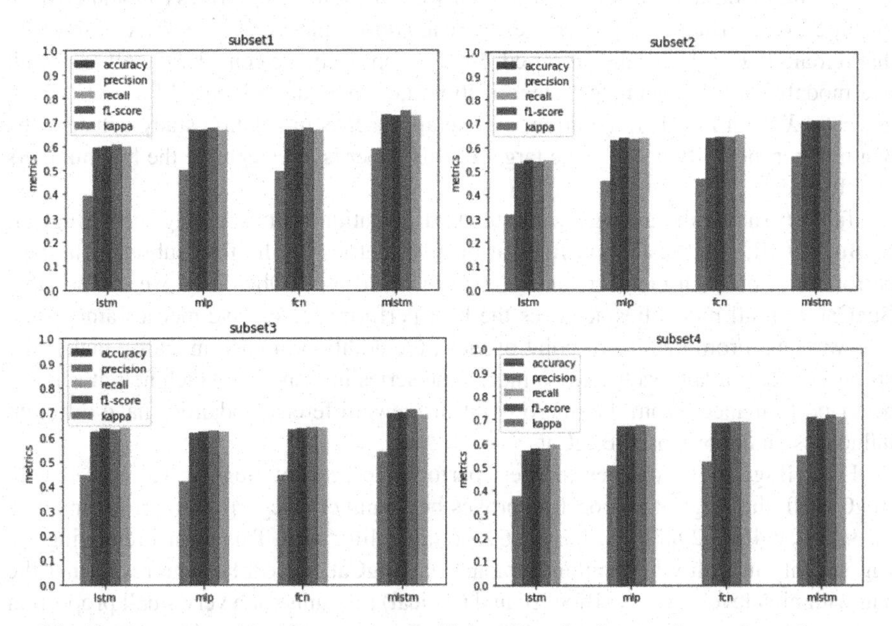

Fig. 5. The results of BugCat compared with baselines for five metrics on four subsets

The better performance of the proposed BugCat approach over three baseline methods prove that this approach can considerably improve bug number categorization prediction. We explain the result that the BugCat consider more modalities in bug tracking system and these modalities contain wealthy information to improve the performance of BugCat approach, while baseline methods that only consider the bug number modality make inadequate use of information in bug tracking system. On the one hand, one modality contains information that other modalities lack, this complementary information among modalities can be used to enhance the model in multi-modal time series learning. For instance, the day of week modality can provide periodic information but the bug severity can't. On the other hand, different modalities can share their information through data fusion. This consistency information can be learned in the training process.

4.3.3 Ablation Study

In order to investigate RQ3, we compare the performance of the proposed BugCat approach that considers all five modalities with four variants of BugCat that consider only four modalities. For the first variant (named BugCat-I), we remove the day of week modality from the inputs of the BugCat approach, i.e., there are only 4 inputs X^{num}, X^{do}, X^{sev} and X^{pri} that remain in Fig. 1. For the second variant (named BugCat-II), we remove the day off modality from the inputs of the BugCat approach, i.e., there are only 4 inputs X^{num}, X^{dw}, X^{sev} and X^{pri} that remain in Fig. 1. For the third variant (named BugCat-III), we remove the bug severity modality from the inputs of the BugCat approach, i.e., there are only 4 inputs X^{num}, X^{dw}, X^{do} and X^{pri} that remain in Fig. 1. For the fourth variant (named BugCat-IV), we remove the bug priority modality from the inputs of the BugCat approach, i.e., there are only 4 inputs X^{num}, X^{dw}, X^{do} and X^{sev} that remain in Fig. 1. For the proposed BugCat approach, we consider the effects of all five modalities on bug number categorization, i.e., there are 5 inputs X^{num}, X^{dw}, X^{do}, X^{sev} and X^{pri} in Fig. 1. In addition, all five approaches in ablation study consider the bug number modality because the target of this paper is to categorize the bug numbers of software.

Table 6 shows the average with standard deviations between BugCat-I, BugCat-II, BugCat-III, BugCat-IV and BugCat on five metrics on the four subsets. The best performance on each metric is indicates in the bold type. Obviously, we can see that BugCat with all modalities achieves the best performance on five metrics among the four variants. From the macro point of view, the addition of each modality makes the proposed BugCat approach contain more time series information which help it achieve better performance. From the micro point of view, different modalities have different influences on the proposed BugCat.

First, BugCat-III achieves the best performance except BugCat. Compared with BugCat-III, the BugCat approach improves bug number categorization on five metrics by 3.42%, 3.46%, 2.88%, 3.21%, 6.66% (relative lift ratio). This means removing the bug severity modality has minimal impact on BugCat, we explain the result that the bug with high level severity (Blocked and Critical) accounts for a very small proportion of the total number of bugs. Second, BugCat-II achieves the second-best performance except BugCat. Compared with BugCat-II, the BugCat approach improves bug number categorization on five metrics by 4.64%, 4.20%, 3.69%, 4.12%, 8.87%. This means removing the day off modality has a bigger impact than removing the bug severity

Table 6. The average values with standard deviations in terms of five metrics of the proposed BugCat approach and four variants on the four subsets.

Model	Accuracy	Macro precision	Macro recall	Macro F-score	Kappa
BugCat-I	0. 6794 ± 0.0115	0. 6906 ± 0.0091	0. 6843 ± 0.0099	0. 6821 ± 0.0105	0. 5158 ± 0.0170
BugCat-II	0. 6801 ± 0.0157	0. 6973 ± 0.0146	0. 6858 ± 0.0170	0. 6848 ± 0.0165	0. 5186 ± 0.0250
BugCat-III	0. 6886 ± 0.0092	0. 7023 ± 0.0086	0. 6912 ± 0.0076	0. 6912 ± 0.0078	0. 5299 ± 0.0158
BugCat-IV	0. 6794 ± 0.0168	0. 6974 ± 0.0189	0. 6814 ± 0.0201	0. 6818 ± 0187	0. 5158 ± 0.0278
BugCat	**0. 7122 ± 0.0115**	**0. 7266 ± 0.0149**	**0. 7111 ± 0.0144**	**0. 7130 ± 0.0137**	**0. 5652 ± 0.0191**

modality. For software developer and end-user, they are more likely to submit bugs on weekdays than on days off and prefer to rest rather than work on their days off. Third, BugCat-IV achieves the third-best performance except BugCat. Compared with BugCat-IV, the BugCat approach improves bug number categorization on five metrics by 4.82%, 4.29%, 4.36%, 4.58%, 9.58%. This means removing the bug priority modality has a bigger impact than removing the bug severity and the day off modalities. We explain the result that the number of bugs with higher priority (P1, P2 and P3) accounts for a larger proportion of the total bug numbers than the number of bugs with higher severity (Blocked and Critical). Finally, BugCat-I achieves the worst performance compared with other variants. Compared with BugCat-I, the BugCat approach improves bug number categorization on five metrics by 5.14%, 5.21%, 3.91%, 4.53%, 9.68%. This means removing the day of week modality has the most impact on BugCat. The submission of software bugs by developers and users are usually affected by periodicity for week, i.e., there are more bugs from Monday to Friday and less bugs from Saturday and Sunday. Ablation study result shows that considering all modalities makes the proposed BugCat approach achieve the best performance, and different modalities have different influences on the proposed BugCat approach. In practice, the project manager should gather as much valuable modality information as possible to predict future bug number categorizations.

5 Conclusion and Future Work

To improve bug number categorization, we propose a novel approach called BugCat with multi-modal time series learning. First, we obtain the time series derived from the five modalities (i.e., the bug number, the day of the week, the day off, bug severity and bug priority) in Mozilla Firefox bug tracking system. Then, the LSTM embedding is conducted on the five modalities of times series separately to capture the time-dependence of each modality. Third, five modalities share information with each other by data fusion and multi-modal time series learning is conducted to categorize the number of bugs in future.

The extensive experiments with the Mozilla Firefox bug data demonstrate the superiority of the proposed BugCat approach over state-of-the-art techniques including multilayer perceptron (MLP), fully convolutional network (FCN) and Long Short-Term Memory (LSTM). In addition, we also apply a greedy strategy to automatically tune the parameters including the hidden units of LSTM embedding, the hidden units of FC layer and batch size. The tuning of parameters makes the proposed BugCat approach to achieve the best performance. Finally, according to the result of the ablation study, we find that, to make improved bug number categorization, the four modalities as the day-of-week, the day-off, bug severity and bug priority are of great importance. Among them, the day off modality is the most important modality.

Although the proposed BugCat approach has demonstrated some promising aspects for improving bug number categorization, we admit that there are still some limitations that demand future work. First, in this paper, we only experimented on four subsets of a Mozilla software project, and we will apply BugCat to more projects. Second, the bug tracking system also contain a wealth of information such as platform, component, and description etc. This information may improve the bug number categorization, and we will incorporate this information into BugCat in future work. Third, the subset division of three categorizations is relatively rough and needs further refinement. We will devise a more objective and reasonable method to divide each subset into three or more categories in our future work.

Acknowledgment. This research is supported in part by Beijing Youth Talent Fund No. Q0011019202001; National Natural Science Foundation of China under Grant No. 72174018; the Beijing Natural Science Foundation under Grant No. 9222001 and the Philosophy and Sociology Science Fund from Beijing Municipal Education Commission (SZ202110005001).

References

1. Chen, X., Zhang, D., Zhao, Y., Cui, Z., Ni, C.: Software defect number prediction: unsupervised vs supervised methods. Inf. Softw. Technol. **106**, 161–181 (2019). https://doi.org/10.1016/j.infsof.2018.10.003
2. Choetkiertikul, M., Dam, H.K., Tran, T., Ghose, A.: Predicting delays in software projects using networked classification (T). In: 2015 30th IEEE/ACM International Conference on Automated Software Engineering (ASE), Lincoln, NE, USA, pp. 353–364. IEEE (2015). https://doi.org/10.1109/ASE.2015.55
3. Runeson, P., Alexandersson, M., Nyholm, O.: Detection of duplicate defect reports using natural language processing. In: 29th International Conference on Software Engineering (ICSE 2007), Minneapolis, MN, USA, pp. 499–510. IEEE (2007). https://doi.org/10.1109/ICSE.2007.32
4. Yang, X., Tang, K., Yao, X.: A learning-to-rank approach to software defect prediction. IEEE Trans. Rel. **64**, 234–246 (2015). https://doi.org/10.1109/TR.2014.2370891
5. Wu, W., Zhang, W., Yang, Y., Wang, Q.: Time series analysis for bug number prediction. In: Proceedings of the 2nd International Conference on Software Engineering and Data Mining, pp. 589–596 (2010)

6. Pati, J., Shukla, K.K.: A comparison of ARIMA, neural network and a hybrid technique for Debian bug number prediction. In: 2014 International Conference on Computer and Communication Technology (ICCCT), Allahabad, India, pp. 47–53. IEEE (2014). https://doi.org/10.1109/ICCCT.2014.7001468
7. Pati, J., Kumar, B., Manjhi, D., Shukla, K.K.: A comparison among ARIMA, BP-NN, and MOGA-NN for software clone evolution prediction. IEEE Access. **5**, 11841–11851 (2017). https://doi.org/10.1109/ACCESS.2017.2707539
8. Zhang, W., Du, Y., Yoshida, T., Wang, Q., Li, X.: SamEn-SVR: using sample entropy and support vector regression for bug number prediction. IET Softw. **12**, 183–189 (2018). https://doi.org/10.1049/iet-sen.2017.0168
9. Andreou, A.S., Chatzis, S.P.: Software defect prediction using doubly stochastic Poisson processes driven by stochastic belief networks. J. Syst. Softw. **122**, 72–82 (2016). https://doi.org/10.1016/j.jss.2016.09.001
10. Rathore, S.S., Kumar, S.: Towards an ensemble based system for predicting the number of software faults. Expert Syst. Appl. **82**, 357–382 (2017). https://doi.org/10.1016/j.eswa.2017.04.014
11. Qiao, L., Li, X., Umer, Q., Guo, P.: Deep learning based software defect prediction. Neurocomputing **385**, 100–110 (2020). https://doi.org/10.1016/j.neucom.2019.11.067
12. Yu, X., Keung, J., Xiao, Y., Feng, S., Li, F., Dai, H.: Predicting the precise number of software defects: Are we there yet? Inf. Softw. Technol. **146**, 106847 (2022). https://doi.org/10.1016/j.infsof.2022.106847
13. Zhang, T., Chen, J., Yang, G., Lee, B., Luo, X.: Towards more accurate severity prediction and fixer recommendation of software bugs. J. Syst. Softw. **117**, 166–184 (2016). https://doi.org/10.1016/j.jss.2016.02.034
14. Tian, Y., Lo, D., Xia, X., Sun, C.: Automated prediction of bug report priority using multi-factor analysis. Empir. Softw. Eng. **20**(5), 1354–1383 (2014). https://doi.org/10.1007/s10664-014-9331-y

Metaheuristic Enhancement with Identified Elite Genes by Machine Learning

Zhenghan Nan, Xiao Wang, and Omar Dib(✉)

Department of Computer Science, Wenzhou-Kean University, Wenzhou, China
{zhenghna,xiaowa,odib}@kean.edu

Abstract. The traveling salesman problem (TSP) is a classic NP-hard problem in combinatorial optimization. Due to its difficulty, heuristic approaches such as hill climbing (HC), variable neighborhood search (VNS), and genetic algorithm (GA) have been applied to solve it as they intelligently explore complex objective space. However, few studies have focused on analyzing the objective space using machine learning to identify elite genes that help designing better optimization approaches. For that, this study aims at extracting knowledge from the objective space using a decision tree model. According to its decision-making basis, a simulated boundary is reproduced to retain elite genes from which any heuristic algorithms can benefit. Those elite genes are then integrated into a traditional VNS, unleashing a remarkable enhanced VNS named genetically modified VNS (GM-VNS). Results show that the performance of GM-VNS surpasses conventional VNS in terms of solutions' quality on various real-world TSP instances.

Keywords: Metaheuristics · Machine learning · Genes selection · Offspring design · NP-hard problems · Elite genes

1 Introduction

Analyzing the structure of the objective space boosts the advancement of metaheuristic optimization simulating natural phenomena as it provides intelligent strategies to explore the search space efficiently. Due to the difficulty of finding a global optimum, metaheuristics have been applied to finding near-optimal solutions for NP-hard problems in a reasonable amount of computational time [11].

Many studies have been conducted to design distinguished algorithms for balancing solutions quality and time complexity. One of the most straightforward approaches is hill climbing (HC), an iterative greedy local search algorithm extensively used in searching problems [12]. However, it has many limitations, especially when the objective space is non-convex; HC tends to get stuck in a local minimum. A simple but effective algorithm is the variable neighborhood search (VNS) designed to handle local minima by systematically exploring distant neighborhood structures [14]. Another metaheuristic is the genetic algorithm (GA), inspired by the natural selection theory. GA has been empirically proven to be an essential candidate for solving large-scale optimization problems in deterministic or stochastic contexts [21]. GAs are robust enough to avoid local minima

J. Chen et al. (Eds.): KSS 2022, CCIS 1592, pp. 34–49, 2022.
https://doi.org/10.1007/978-981-19-3610-4_3

by using advanced evolutionary strategies such as selection, crossover, and mutations. Both VNS and GA have shown exemplary performance in practice and can evolve with hybridization [7]. Although their application and theory are solid, the research on the influence of distinctive patterns(genes) over metaheuristics is deficient. More specifically, in the process of both GA and VNS, there have been few endeavors to identify significant genes and retain their positive impact on offspring. This paper aims to study whether machine learning models can learn from the structure of an objective space to identify substantial genes as the guidance of metaheuristics enhancement. This paper considers the optimization version of TSP as an example of an NP-hard problem over which the proposed method will be tested. Informally speaking, given a complete graph on n vertices and a weight function defined on the edges, the objective of the TSP is to construct a tour (a circuit that passes through each vertex exactly once) of minimum total weight. The TSP is an example of a hard combinatorial optimization problem; the decision version of the problem is NP-complete. The TSP can be formulated as an integer linear programming problem, as shown in [23].

This study hypothesizes that prominent genes identified from random and local minima solutions are conducive to other metaheuristics on exploration quality. Those unique genes stand out regardless of the host algorithm and lead to better searchability. Firstly, this study generates training data for the decision tree with HC. Secondly, it visualizes the model to study how the decision tree identifies excellent genes. Thirdly, it reproduces the decision-making boundary to guide metaheuristics design based on the tree decision-making. Fourthly, it designs the elite genes conservation generator to insert significant genes into the random solutions as the input of enhanced VNS. Fifthly, it develops the enhanced VNS according to the strong genes' conservation idea. Finally, it compares the traditional and improved VNS to inspire the design of a superior GA.

2 Literature Review

As the fundamental local search metaheuristics, 2-OPT and 3-OPT with the amalgamation of other existing metaheuristics have been widely applied to efficiently tackle NP-hard routing problems such as travelling salesman problem (TSP) and vehicle routing problem (VRP). For example, [2] implemented a hybridization of antlion optimization (ALO) and a 2-OPT algorithm to solve the multi-depot vehicle routing problem (MDVRP) by the extensibility of the 2-OPT algorithm. The results present that the hybridization of metaheuristics outperforms the discrete ALO, GA, and ACO in most cases, proving that the particular extensible attribute is conducive to various traditional metaheuristics. Besides the extensibility, 2-OPT performs well in practice due to its simplicity and fast convergence. For instance, [10] conducted robust research for finding approximate solutions to the TSP, and demonstrated that 2-OPT produces a better solution than christofides' algorithm. However, 2-OPT, as a primary local search, turns out to perform poorly in a complex landscape containing so many local minima. Thus, many researchers focused on either improving other metaheuristics based on 2-OPT or embodying HC strengths into other metaheuristics through hybridization. For example, [6] successfully extracted knowledge from 2-OPT via deep reinforcement learning, promoting a better exploration efficiency. Their study reflects that by exploiting 2-OPT, knowledge about the problem structure can be extracted improving the optimization efficiency. Similarly, this paper researches and takes advantage of the particular compositions of 2-OPT and three similar neighborhood structures, 1-OPT, 3-OPT, and city insertion, as the basis of hybridization in VNS.

With the excellent extensibility of the neighborhood structure, as mentioned before, hill climbing, an iterative local search technique, will empower them in light of finding the local optima solution in the current search region. HC is one of the most general heuristics since it has low operational and time complexity to guide other methods such as evolutionary algorithms efficiently. For example, [16] proposed a novel approach based on HC to control the coupling between the transmitter and a receiver in a telecommunication context. They indicated that HC could produce a much better output than the initial solution. In addition, they argued that HC is cost-effective, giving the system a fast response. In [13], the authors proposed an innovative hybridization between HC and particle swarm optimization (PSO), whose results proved the superiority of hybridization. However, HC can quickly get stuck in local minima, as [1] claimed. Thus, they applied an advanced version of hill climbing named β hill climbing, balancing exploration and exploitation.

This paper proposes a novel elite genes conservation approach based on variable neighborhood search (VNS), as the HC approach tends to get stuck into local optima. VNS algorithm was mainly proposed to handle the local minima issues. VNS switches the neighborhood structure via configurable and flexible shaking and local search functions. For example, in [4], the authors significantly produced better solutions for the wind farm layout optimization problem by combining different shake strategies in VNS, an instructive idea inspiring this paper.

What's more, the flexibility of VNS can prevent the search from falling into local minima, which has been verified by [17] on solving multi-depot green vehicle routing problem. Their result proved that a higher quality solution could be obtained by using multiple neighborhood frameworks to jump out of the basin of a particular search region. Although VNS can effectively handle local minima and combine a variety of neighbor structure search algorithms to find the global optima, there is still a mass of randomness in its search process. Tackling this randomness has resulted in an effective strategy for solving the multiperiod inventory routing problem (IRP), as shown in [8]. The authors used an initial fit solution to avoid unnecessary search iterations and promote VNS performance in their algorithm. Different from their work, this study applies a different technique, namely elite genes conservation, to interiorly dwindle the impact of randomness and intelligently guide the search process.

Both VNS and GA are powerful metaheuristics effectively escaping local minima. Interestingly, the GA imitates a biological evolution process inspired by Darwin's natural evolution theory for its computational model. GA has strong operability since it can directly operate on sets, sequences, matrices, trees, graphs, and other structural objects with different encoding strategies. For example, in [3], the authors studied solving community detection problem (CDP) with GA, and they applied edge-based encoding to represent individuals being evaluated by fitness function straightforwardly. GA is scalable and can be efficiently run in parallel and effectively combined with other algorithms. For instance, [20] conducted a study on combining convolutional neural networks (CNN) with GA for image classification, and they obtained a remarkable improvement in the CIFAR image dataset. In addition, [15] proposed a comprehensive optimization of engine efficiency by coupling artificial neural network (ANN) with GA.

Nonetheless, the mutation in the GA increases the uncertainty of the result. One strategy for improving GA is avoiding blind mutation. [9] adopted that method by proposing a time-varying mutation operator. Compared with traditional GA, using a dynamic mutation operator improves performance. Based on their research, our work points out one of the shortcomings of conventional GA mutation. It puts forward a novel approach to mutating intelligently based on the identified elite genes.

The algorithms above advance significantly with machine learning (ML). Many studies showed that ML has solid performance in extracting knowledge from data. For example, [18] showed that knowledge extraction by ML algorithm promotes the future development of materials science. The ML techniques are widely applied in theory and practice. For instance, [22] argued that combining hybrid metaheuristics, data envelopment analysis, and k-mean clustering is successful for extracting knowledge from metaheuristic algorithms. As a result, extracting knowledge from solutions' structures of metaheuristic algorithms with ML is feasible. This idea has been proved by [20] by classifying genes for early cancer detection and prevention. They reduced the search space and the complexity of the metaheuristic algorithm with ML and data mining.

Moreover, genetic engineering endows individuals with excellent traits by strengthening some genes. For example, [5] mentioned that plant genetic engineering improves crop productivity by giving crops ideal genetic characteristics. Equivalently, identifying elite genes via ML models and maintaining those positive traits to avoid being destroyed during the offspring's optimization will potentially help create strong offspring and lead to faster convergence. Inspired by those ideas, our work applies ML techniques to analyze the structure of objective space, breeding an astute space exploration policy.

3 Methods

This research studies whether a machine learning model can extract useful information from the objective space of TSP, an NP-hard problem in combinatorial optimization. The acquired knowledge from the machine learning model will be used to identifying elite genes with which an intelligent solutions generator and a smart intervention strategy are implemented to enhance existing optimization methods. This work also shows that all the studied algorithms have benefited from the outstanding genes resulting from the ML model; such genes have considerably improved the quality of obtained solutions. As a research methodology, the conventional concepts of VNS and neighborhood structures will be introduced for the convenience of following enhanced algorithms explanation. Secondly, we discuss how the training data of the decision tree model is represented, generated and labeled. Thirdly, the source and representations of instances will be explained. Fourthly, the process of training and visualizing the decision tree will be illustrated. Fifthly, the simulated decision boundary will be reproduced so that the elite genes can be identified. Finally, genetically modified variable neighborhood search (GM-VNS) will be designed based on the elite genes mimicking the simulated decision boundary. In the following, we detail the experimental process.

3.1 Traditional Variable Neighborhood Search (VNS)

X-OPT and city insertion have been extensively studied for the TSP problem. The core idea of 1-OPT is to select an interval for optimization randomly. This algorithm randomly selects one city from the TSP circuit and exchanges it with every other city in the solution to generate the list of neighbor solutions. The process moves to the best neighbor and repeats until a local minimum is encountered. Similar to 1-OPT, 2-OPT and 3-OPT, delete the connection between two and three non-adjacent nodes in the TSP circuit, try other connection modes, calculate the path length after different connection modes, and select the connection mode with the shortest path length as the new solution. Similarly, the city insertion process chooses one element and inserts it into two adjacent cities. Then the process is repeated until a local minimum is encountered. For more information about the adopted neighborhood structures, please refer to [19].

In HC, the local search keeps generating new solutions based on the current solution and comparisons with the current best solution. The latter is updated if the new solution is better than the best. If not, the iteration stops and the best solution is returned. VNS successfully explores a set of predefined neighborhoods to provide a better solution. The algorithm proceeds with an initial solution randomly generated. Next, based on the initial solution and a neighborhood structure, VNS applies a shaking operation that attempts to jump out of the current local minimum region while making the local optima closer to the global optima. After shaking, VNS applies an HC to find the local optima and improve the search accuracy. After HC, the algorithm checks if the resulting solution is improved compared to the best solution. If so, the local minimum will be the new initial solution and search is repeated. If not, the algorithm switches the neighborhood structure and generates a new solution in this neighborhood, and improves it.

3.2 Training Data Generation

As discussed before, this work applies machine learning to analyze the structure of solutions' variables. The data consists therefore of a set of solutions satisfying the requirements of TSP. As for the data features, we use all the edges of a given TSP instance to denote the features. That is, each feature represents an edge, and its value is either 0 or 1. 1 means that the edge is selected to construct the solution (i.e., TSP circuit), whereas 0 means that the edge does not appear in the solution. For example, for a TSP instance with n cities, there exists $(n * (n - 1))/2$ edges |E| corresponding to the number of data features. In order to satisfy the TSP requirements for every data sample, among all the features, only a subset n of |E| will have a value of 1, and the rest of features are set to 0. For the labels, each sample is labeled based upon two values, either 0 or 1. 1 indicates that the sample is a local minimum with relatively higher quality resulting from a local search procedure such as hill climbing, whereas 0 means that the solution is randomly generated, hence of lower quality. The aim is to make two clusters of relatively good and bad solutions, and subsequently analyze the behavior of the decision tree that classifies those solutions. Analyzing the decision boundary helps identifying specific common features in good and bad solutions. Algorithm 1 shows the pseudo-code of the training data generator.

Algorithm 1 Trainning Data Generator

```
Input: trainningDataSize n
Output: List<String>trainningData data
 1: InstanceSize ← k
 2: for i 0 to n by 1 do
 3:     s1 ← randomSolutionGenerator(k)
 4:     s2 ← HCwith2OPT(s1)
 5:     if s2.Betterthan(s1) then
 6:         label ← 1
 7:     else
 8:         label ← 0
 9:     end if
10:     newRecord ← features(edges) + label
11:     data ← data + newRecord
12: end for
13: return data
```

To generate the dataset without bias, a generator randomly selects one city each time and inserts it into the final solution if the selected one isn't repetitive. When all positions of the solution are filled, the generator stops. The random solution of relatively low quality is labeled to 0. As for the local search algorithm used to generate data with label 1, the hill climbing algorithm with 2-OPT, a simple local search algorithm for solving the TSP, is responsible for improving a random solution. There are four neighborhood structures implemented in the experiments, but only 2-OPT is selected as the specification of HC in the data generation phase. The reason is that 2-OPT is known to have the best performance (i.e., produce the best local minima) among the four structures when being applied with HC. An empirical study has been conducted to support that statement, and its results are presented in Table 1.

Table 1. Comparison of HC with different neighborhood structures

Metric\Algorithm	HC-1OPT	HC-2OPT	HC-3OPT	HC-CI
Average distance for TSP (KM)	2501.386	2085.822	2718.354	2422.298

According to Table 1, HC-2OPT has the shortest distance as fitness for the TSP problem compared with other neighborhood structures, namely 2-OPT, 3OPT and city insertion. Therefore, HC-2OPT is selected to generate the relatively higher quality solutions that will have the label 1. We believe that random and higher-quality solutions (i.e., local minima resulting from HC-2OPT) comprise useful information that helps identifying elite genes. This assumption will be verified during the experiments later.

3.3 Test Instances

The distance information of a TSP instance is stored in the form of a matrix specifying the distance value between every pair of cities. Four real-world TSP instances are used in this work: bays29, berlin52, eil76, and eil101. The suffix indicates the number of cities in the instance, so the larger the number is, the more complicated the NP-hard problem will potentially be. Table 2 shows the instances with their descriptions.

Table 2. Description of the selected instances

	Bays29	Berlin52	Eil76	Eil101
Source description	29 Cities in Bavaria (street distance)	52 locations in Berlin (Germany) (Groetschel)	76-city problem (Christofides/Eilon)	101-city problem (Christofides/Eilon)
Optimum tour (km)	2020	7526	538	629

To study the impact of the problem structure, we also use other instances obtained using a simple TSP random generator. The random instance generator can control the instance's range, mean, and variance of distances.

3.4 Visualize the Decision Tree Model to Identify Elite Genes

In the following sections, the decision tree algorithm is applied to the generated datasets to analyze the decision-making criteria and check if it is highly explanatory, facilitating the simulation of decision-making boundaries. The results present that the accuracy of the decision tree model is always very high (>98%) in all the studied instances in this paper. Consequently, it reveals a clear boundary between the local minimum and random solutions. Interestingly, the model has identified some patterns in the data and, therefore, learned from the TSP solution space. Due to substantial decision criteria, the decision visualization is done to study how the decision tree classifies local minimum and random solutions based on several specific edges. According to the visualization, the paths to classify local minima are followed to determine which edges will be fixed and removed. By retaining significant edges, those essential features of outstanding solutions (i.e., local minima) can exert their advantages to optimize future solutions' generators and optimization methods. We refer to those edges as elite genes.

3.5 Genetically Modified Variable Neighborhood Search (GM-VNS)

Retaining outstanding genes (i.e., specific edges in a TSP) by binding two ends (i.e., cities) of edges in VNS is the basic idea of genetically modified variable neighborhood search (GM-VNS). We aim to check whether the identified elite edges can be exploited to improve the performance of a local search procedure such as VNS. The method proceeds with an initial solution generated based on the simulated decision-making resulting from the decision tree classification model. Then locate and package elite genes in the solution in light of simulated decision-making boundary. After the solution is packaged, in the shaking and local search processes, the operations will be based on a set of pre-assembled elite genes rather than single cities as in a traditional VNS. The process is equivalent to manipulating a sequence of pre-identified fit edges instead of manipulating one single city. Following is the flow chart (see Fig. 1) of GM-VNS.

The main difference between traditional VNS and GM-VNS is that GM-VNS operates based on a block of fixed cities rather than single ones. In GM-VNS, the very first

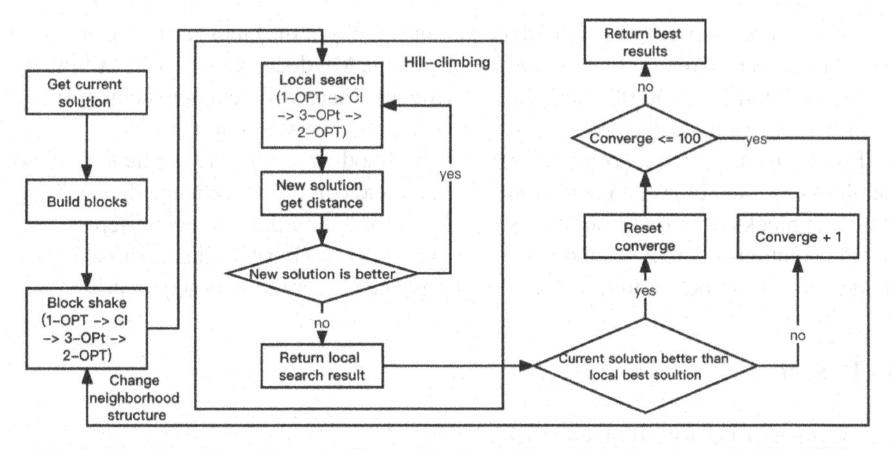

Fig. 1. The flow chart of genetically modified variable neighborhood search (GM-VNS).

operation is block-building. For a TSP problem, traditional local search algorithms use single cities as fundamental elements to change the solution structures. In GM-VNS, a procedure will build blocks of fixed cities before applying local search. That is done by adding the cities of elite edges into an array according to the simulated decision-making basis. A rule list is followed to determine which city should be bound and compute each block's size and elements. It is worth noting that the incumbent solution of GM-VNS already meets the binding requirement by making binding cities adjacent.

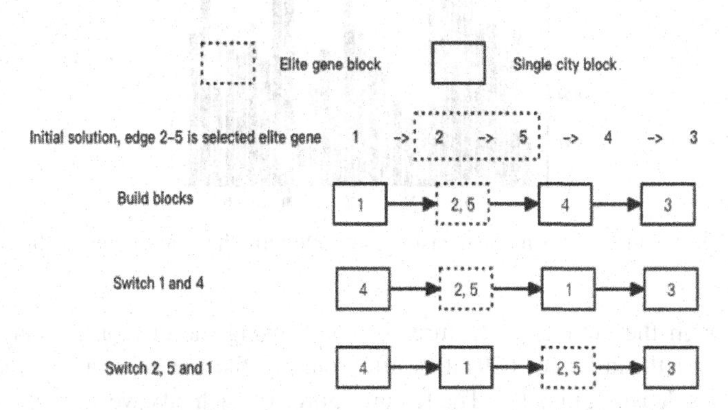

Fig. 2. Illustration of integrating elite genes into VNS

Since the data structure to represent solutions is an array, the block representing solutions is also converted to an array. If the current city should be bound, first, GM-VNS will get its partners and add them to the following block; if not, only the current city will be added to a block with one element. After building blocks, they will be loaded into an array list as a block list. Figure 2 illustrates the concept of blocks and the swap operations underlying it. As can be seen the cities 2 and 5 operates as a single element.

The shaking process in traditional VNS generates a random neighbor solution in the current neighbor structure based on a single element by swapping cities' positions. In

GM-VNS, the swapping operation is based on the blocks of the block list as illustrated in Fig. 2. The aim is to retain some edges based upon the ML decision boundary by binding cities on their sides. After the new shaking strategy, there will be a new neighborhood solution having some edges whose presence will make the solution more fit.

For the local search algorithm, four neighborhood structures are applied based on the blocks of fixed cities and not on single cities. Such algorithms change the positions of certain blocks as in traditional VNS. By doing so, during the HC, every step ensures that the solutions contain those fixed outstanding edges. After HC, a comparison helps update the current best solution. After iterations, the best solution is acquired.

4 Result

4.1 Impact of Dataset Representation

This section studies the impact of data representation on model performance. As mentioned before, the edges represent the features of the labeled datasets in our model. However, one could also use the cities instead of the edges for the features. In that case, the value of each feature represents the order of the city in the TSP circuit.

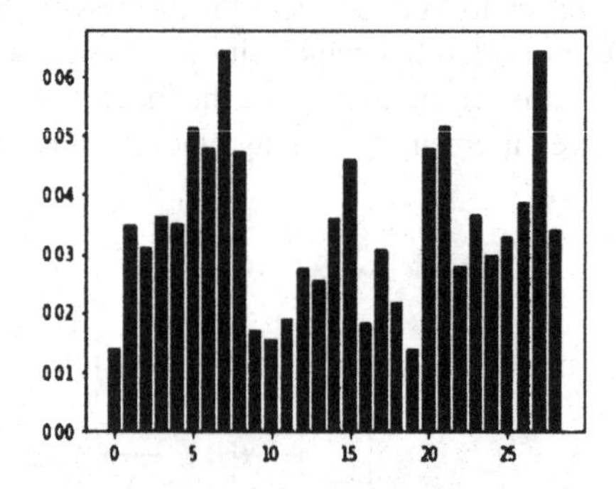

Fig. 3. Feature scores in the decision tree following the "city representation".

Even though the number of features is small, using the city order representation showed that with the instance of 29 cities, after training, the result was a vast tree with too many branches, leaves and paths. The feature scores of each city were similar, meaning that it's difficult to reproduce the simulated decision boundary guiding GM-VNS since each feature has comparable importance. In other words, the city order representation does not help extract knowledge from the objective space effectively. Following are the feature scores (see Fig. 3) in the tree above.

Due to the inefficiency of city order representation, edge representation was adopted. When the features are edges rather than the city index, after training, the tree becomes more insightful (see Fig. 4). The edge representation leads to a more concise tree and has an explicit feature importance difference, meaning that the ML model can effortlessly identify outstanding genes. From a spatial analysis point of view, edge representation outperforms the city order because the objective space constituted seems to be a near

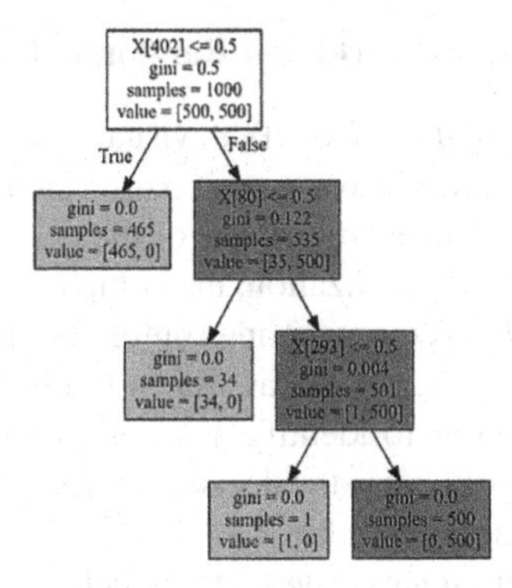

Fig. 4. The visualization of the decision tree following "edge representation".

convex landscape. In contrast, the one constructed by the city order appears to contain many ridges and basins. Naturally, a complex landscape hinders the ML model from identifying decisive factors qualifying the solutions. Hence, the ML model will not generalize well. Interestingly, in the instance with 29 cites, edges 402, 80, and 293 are the most important genes, and they are conserved for the following algorithms.

4.2 Decision Tree Visualization Analysis

Based on the finding that there is a robust decision boundary between local minima and randomly generated solutions, the boundary location is researched in the following sections. Excellent features can be passed on to the next generation by reproducing the boundary to guide solutions' creation. Concerning the edge selection, analyzing instance Bays29 is used to illustrate the process of reproducing decision boundaries.

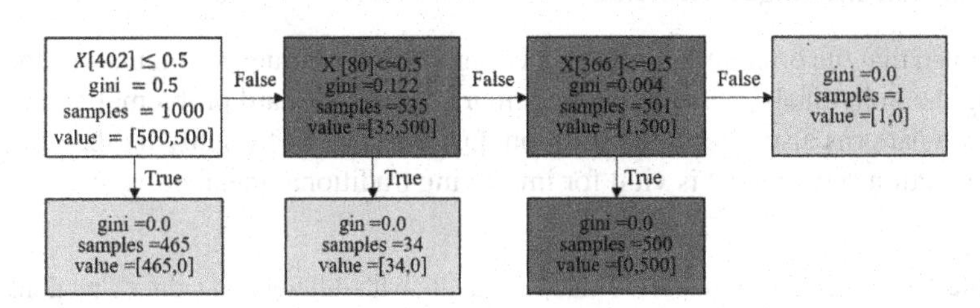

Fig. 5. Decision Tree for Bays29 (Color figure online)

According to Fig. 5, the orange nodes represent random solutions, whereas the blue ones are local minima. In the first node, X indicates the list of features in training data, which corresponds to total edges. It's worth noting that the features' value is either 0 or 1, as introduced before, and the edge is selected only when the value is 1. Hence, $X[402] <= 0.5$ represents a decision on the selection of the edge with index 402. As can be seen, the edge with index 402 is selected to construct the final high-quality solution. Thus, the selected edges inferred from this image are 402 and 80. And the one to be removed is 366. For city nodes binding, the selected edges are converted into a connected city index in terms of the decision.

4.3 Comparison Between Real World and Randomly Generated Instances

Interestingly, with the training data size of 1000, visual analysis shows that the decision trees from random instances are always more complicated (i.e., more leaves, nodes, paths, etc.) than real-world instances, regardless of the degree of randomness involved and instance size. In light of the visualization, the complexity of the tree doesn't vary a lot, even though there is a vital difference among configurations. By contrast, the decision tree of real-world data is always as simple as shown in Fig. 5. In other words, the ML algorithm has more opportunities to identify a set of elite edges. This is an essential characteristic because the more concise the decision-making basis, the more feasible the reproduction of decision boundary will be.

On the contrary, a tree with many leaves and branches disturbs decision-making as each branch might represent a specific local minimum region equivalent to significant genes. The divergence of those genes in the search space triggers the difficulty of integrating their advantages into a single expected solution. That is, a concise tree structure is preferable to keep the eliteness of training data. For real-world instances, the number of clusters containing solutions with outstanding genes is relatively small. It can be described as a flat field with two near basins with unique characteristics that make them shine. That corresponds to a decision tree with few paths (basins) with a particular number of genes. Such a near-convex objective space is conducive to the model's learning. In contrast, the random instances can be seen as a field with so many basins next to each other, which are different and not that deep. That corresponds to a solution space with so many ridges or plateaux. Naturally, learning from such spaces is a difficult task as the existence of many holes with different genes. As such, the ML algorithm will be deficient in generalization ability.

4.4 Impact of the Edges Direction

It's accepted that the orientation of the genes significantly influences their surroundings, affecting the individuals' performance. Similarly, in an NP-hard problem, the direction of solution patterns also affects the solution quality. That's why studying the impact of edges' direction (i.e., genes) is vital for improving traditional metaheuristics.

Table 3. Comparing the impact to solution quality with different bounding directions.

	Metric\Instance	Bays29	Berlin52	Eil76	Eil101
Traditional VNS	Average distance (km)	2053.89	2054.18	2051.44	2052.81
	Average time (ms)	27.404	27.432	27.609	28.196
GM-VNS	Average distance (km)	2046.21	2064.08	2055.42	2068.03
	Average time (ms)	31.914	31.103	32.583	30.94

Like genes, the direction critically influences the surroundings. An experiment was conducted to determine the importance of the edges' direction in solutions. In the instance of 29 cities with 406 edges, according to the simulated decision-making basis, cities "21" and "13", "2" and "28", "16" and "17" are fixed. They were then integrated into GM-VNS to perform the optimization and compared with the solutions' quality of traditional VNS using the same instance but no bound cities. The running time and solution quality of the four testing groups are shown in Table 3.

The comparison results show that the quality of solutions from GM-VNS is different when changing the bounding order in the fixed blocks, proving that the direction of edges impacts the solutions' quality. Likewise, in the biology field, the orientation of adjacent genes has been reported to influence their expression in eukaryotic systems, a convincing analogy fortifying the impact of elite genes orientation on the surroundings.

Therefore, it's essential to find an appropriate bounding direction for adjacent cities, rendering an enhanced performance during the process of exploration and exploitation. Testing the performance in different directions indicates that the best configuration for the 29 cities instance (Bays29) is "21,13", "2,28" and "16,17".

4.5 Conservation Generator Performance Assessment

By applying a simulated decision-making basis, significant genes will be preserved as much as possible as the foundation of the enhanced generator. This subsection aims to study the impact of the quality of the initial solution on the optimization process.

Table 4. Comparison between a traditional and enhanced generator based on elite genes

Metric\Instance	Bays29	Berlin52	Eil76	Eil101
Traditional generator distance (km)	5971.434	29869.411	2501.536	3389.889
Enhanced generator distance (km)	5498.875	28012.23	2384.882	3313.372
Improvement from traditional to enhanced generator in (km)	472.559	1857.181	116.654	76.517
Improvement from VNS to GM-VNS in (km)	13.299	90.469	3.543	2.325

Table 4 presents a substantial enhancement in favor of the assumption that significant genes play an essential role in constructing high-quality solutions. The third row in Table 4 shows that integrating elite genes into the generator results in better initial solutions. By comparing the improvement extent between a traditional VNS with a classical generator and GM-VNS with an improved generator, the fourth row in Table 4 shows that GM-VNS performs better. Therefore, random solutions are easier to be improved than local optima. This phenomenon is apprehensible because the random solutions diverse in objective space contribute to a comprehensive scope searching, but local minimum solutions are usually concentrated in clusters in space. Thus, the impact of outstanding genes diminishes gradually with the solutions' evolution.

Table 5. Comparison of VNS and GM-VNS based on elite genes (average of 200 simulations)

Metric\Instance	Bays29	Berlin52	Eil76	Eil101
VNS distance (km)	2055.998	8018.654	554.597	644.729
GM-VNS distance (km)	2042.699	7928.185	551.054	642.404
Improvement (km)	13.299	90.469	3.543	2.325
VNS convergence time (ms)	37.407	420.523	1826.019	6389.356
GM-VNS converge time (ms)	43.473	535.943	2850.11	8580.196
Time difference	6.066	115.42	1024.091	2190.84

4.6 VNS and GM-VNS Comparison Results

As can be seen from Table 5, GM-VNS always results in better quality solutions than traditional VNS across all four instances. Results demonstrate that retaining outstanding genes did improve the quality of solutions, including local minima. In other words, these originally unobtrusive genes play a significant role, especially when the traditional metaheuristic reaches its performance ceiling. What's more interesting is that the identification of significant genes is based on the local minimum from hill climbing.

However, the experience from HC, a relatively weak algorithm, can enhance VNS performing much better than HC. Therefore, excellent genes are not dependent on a specific algorithm. All the metaheuristics can benefit from them but with different configurations and improvements. The improvement does not necessarily appear to be significant as the increase in instance size. For instance, the improvement of Bays29 is 13.299, and Berlin52 is 90.469, which is much bigger than Bays29. Although Eil76 has a larger instance size than Berlin52, its improvement is 3.543 more minor than Bays29. Therefore, the improvement extent is determined by instance size and other factors such as the instance's distribution and objective landscape. As for the time analysis, results demonstrate that GM-VNS needs a relatively more or similar time to converge compared with VNS. The convergence time is also influenced by the implementation and the additional time to retain and identify dominant genes. As the increase of instance sizes, the time difference becomes more prominent because the NP-hard problems become much more complicated, leading to an increased disadvantage.

4.7 Impact of Elite Genes on Genetic Algorithm

Metaheuristics can perform well with elite genes with different configurations. However, GA might undo the impact of elite genes due to its separate evolution process, which is an interesting direction to improve GA in the future. The defects of traditional GA will be revealed through the following experiment to determine if the strategy of fixing edges in initial solutions enhances the performance of GA. Four groups applied different instances with 29, 52, 76, and 101 cities. For each group, we compare the performance of GA with untreated genes and elite genes. The following are the results of four groups of experiments (average of 200 test data) (see Table 6).

Table 6. Comparison of GA with untreated gene and fixed gene

	Metric\Instance	Bays29	Berlin52	Eil76	Eil101
Traditional GA	Average distance (km)	2080.21	8157.9	594.71	731.40
	Average time (ms)	445.01	1264.92	2227.2	3554.10
Enhanced GA	Average distance (km)	2084.65	8141.28	597.98	736.17
	Average time (ms)	433.87	1290.145	2130.04	3384.18

The difference between GA with untreated genes and fixed genes is slight. In GA, there are processes of blind mutation and crossover, which eventually destroy the fixed edges. Consequently, the result will be similar to the process of using a GA with untreated genes. The results show that the solution using selected genes has lower quality than the solution using untreated genes for the instance Bays29. Nevertheless, the result was reversed in the group using the instance Berlin52. Since the traditional GA mechanisms prevent elite genes from optimizing the internal search process, the quality difference between conventional and improved versions remains negligible. Therefore, the enhanced input solutions injected with elite genes do not benefit traditional GA because the internal evolution process broke and separated those elite genes. Consequently, it's necessary to adapt the genetic operations of the GA to be consistent with the idea of conserving elite genes in the future.

5 Conclusion

Since the conventional VNS and GA neglect specific genes' domination, the systematical evolution process appears indiscriminate. Therefore, it's essential to understand how elite genes guide the evolution of offspring and how they interact with their surroundings. By implementing a genetically modified VNS, this study showed that using the edge representation in the training data is more efficient than the city order representation. The edge representation is beneficial to model learning and generalization.

This study showed that tackling real-world instances differs from random instances from an objective space exploration point of view. Results indicated that real-world instances tend to have a near-convex objective space, whereas the random ones' local optima scatter sporadically in space, impeding the model's learning process. In addition, the genes' direction influence in metaheuristics is predominant, which is persuasive evidence of the interaction between genes and their surrounding environment. Empirical analysis showed that GM-VNS surpasses VNS in solutions quality, verifying those elite genes can strengthen solutions' quality regardless of the algorithm types.

Further research on the genes' environmental interaction and how to take advantage of those elite genes should focus on the following directions. Researchers can develop augmented mutation or mating for GA by dynamically injecting elite genes to avoid blind genetic operators' impact. Other metaheuristics can benefit from elite genes too. Thus, those enhanced metaheuristics are worthwhile to implement. Although the proposed method was only applied to TSP, it is relevant to check its performance on other

NP-hard problems such as knapsack. It's valuable to try different training data representations, contributing to a more effective decision-making boundary. Finally, other machine learning models may reveal more information about elite genes.

References

1. Al-Betar, M.A., Awadallah, M.A., Abu Doush, I., Alsukhni, E., ALkhraisat, H.: A non-convex economic dispatch problem with valve loading effect using a new modified β-Hill climbing local search algorithm. Arab. J. Sci. Eng. **43**(12), 7439–7456 (2018)
2. Barma, P.S., Dutta, J., Mukherjee, A.: A 2-opt guided discrete antlion optimization algorithm for multi-depot vehicle routing problem. Decis. Mak. Appl. Manage. Eng. **2**(2), 112–125 (2019)
3. Bello-Orgaz, G., Salcedo-Sanz, S., Camacho, D.: A multi-objective genetic algorithm for overlapping community detection based on edge encoding. Inf. Sci. **462**, 290–314 (2018)
4. Cazzaro, D., Pisinger, D.: Variable neighborhood search for large offshore wind farm layout optimization. Comput. Oper. Res. **138**, 105588 (2022)
5. Cunningham, F.J., Goh, N.S., Demirer, G.S., Matos, J.L., Landry, M.P.: Nanoparticle-mediated delivery towards advancing plant genetic engineering. Trends Biotechnol. **36**(9), 882–897 (2018)
6. Da Costa, P.R.D.O., Rhuggenaath, J., Zhang, Y., Akcay, A.: Learning 2-opt heuristics for the traveling salesman problem via deep reinforcement learning. arXiv preprint arXiv:2004.01608 (2020)
7. Dib, O., Moalic, L., Manier, M.A., Caminada, A.: An advanced GA–VNS combination for multicriteria route planning in public transit networks. Expert Syst. Appl. **72**, 67–82 (2017)
8. Gruler, A., Panadero, J., de Armas, J., Pérez, J.A.M., Juan, A.A.: A variable neighborhood search simheuristic for the multiperiod inventory routing problem with stochastic demands. Int. Trans. Oper. Res. **27**(1), 314–335 (2020)
9. Hasan, M.M., Kashem, M.A., Islam, M.J., Hossain, M.Z.: A time-varying mutation operator for balancing the exploration and exploitation behaviours of genetic algorithm. In: 2021 3rd International Conference on Electrical & Electronic Engineering (ICEEE), pp. 165–168. IEEE, December 2021
10. Hougardy, S., Zaiser, F., Zhong, X.: The approximation ratio of the 2-Opt Heuristic for the metric Traveling Salesman Problem. Oper. Res. Lett. **48**(4), 401–404 (2020)
11. Hussain, K., Mohd Salleh, M.N., Cheng, S., Shi, Y.: Metaheuristic research: a comprehensive survey. Artif. Intell. Rev. **52**(4), 2191–2233 (2018). https://doi.org/10.1007/s10462-017-9605-z
12. Jately, V., Azzopardi, B., Joshi, J., Sharma, A., Arora, S.: Experimental analysis of hill-climbing MPPT algorithms under low irradiance levels. Renew. Sustain. Energy Rev. **150**, 111467 (2021)
13. Kato, E.R.R., de Aguiar Aranha, G.D., Tsunaki, R.H.: A new approach to solve the flexible job shop problem based on a hybrid particle swarm optimization and Random-Restart Hill Climbing. Comput. Ind. Eng. **125**, 178–189 (2018)
14. Kong, M., Xu, J., Zhang, T., Lu, S., Fang, C., Mladenovic, N.: Energy-efficient rescheduling with time-of-use energy cost: application of variable neighborhood search algorithm. Comput. Ind. Eng. **156**, 107286 (2021)
15. Li, Y., Jia, M., Han, X., Bai, X.S.: Towards a comprehensive optimization of engine efficiency and emissions by coupling artificial neural network (ANN) with genetic algorithm (GA). Energy **225**, 120331 (2021)

16. Rohan, A., Rabah, M., Talha, M., Kim, S.H.: Development of intelligent drone battery charging system based on wireless power transmission using hill climbing algorithm. Appl. Syst. Innov. **1**(4), 44 (2018)
17. Sadati, M.E.H., Çatay, B.: A hybrid variable neighborhood search approach for the multi-depot green vehicle routing problem. Transp. Res. Part E Logist. Transp. Rev. **149**, 102293 (2021)
18. Schleder, G.R., Padilha, A.C., Acosta, C.M., Costa, M., Fazzio, A.: From DFT to machine learning: recent approaches to materials science–a review. J. Phys. Mater. **2**(3), 032001 (2019)
19. Sengupta, L., Mariescu-Istodor, R., Fränti, P.: Which local search operator works best for the open-loop TSP? Appl. Sci. **9**(19), 3985 (2019)
20. Sharma, A., Rani, R.: C-HMOSHSSA: gene selection for cancer classification using multi-objective meta-heuristic and machine learning methods. Comput. Methods Programs Biomed. **178**, 219–235 (2019)
21. Sun, Y., Xue, B., Zhang, M., Yen, G.G., Lv, J.: Automatically designing CNN architectures using the genetic algorithm for image classification. IEEE Trans. Cybernet. **50**(9), 3840–3854 (2020)
22. Tayal, A., Solanki, A., Singh, S.P.: Integrated framework for identifying sustainable manufacturing layouts based on big data, machine learning, meta-heuristic, and data envelopment analysis. Sustain. Cities Soc. **62**, 102383 (2020)
23. Rokbani, N., et al.: Bi-heuristic ant colony optimization-based approaches for traveling salesman problem. Soft. Comput. **25**(5), 3775–3794 (2020). https://doi.org/10.1007/s00500-020-05406-5

An End-to-end Weakly-supervised News Aggregation Framework

Xiaohui Huang[1,2] and Xijin Tang[1,2](✉)

[1] Academy of Mathematics and Systems Science, Chinese Academy of Sciences, Beijing 100190, China
huangxiaohui@amss.ac.cn, xjtang@iss.ac.cn
[2] University of Chinese Academy of Sciences, Beijing 100049, China

Abstract. On the Internet era, there are plenty of important events hidden in the mass of news media. Using automated tools to aggregate the valuable media news relative to those events is meaningful for specific media data. In this paper, we present an end-to-end weakly-supervised news aggregation framework that enables tracking the evolution of topics from news. The framework combines Snorkel-based weakly-supervised classification, Latent Dirichlet Allocation (LDA) topic modeling, and topic signal detection model to classify and aggregate unlabeled news texts and ultimately generate visualized results containing news categories, news topics, and temporal topic relationships. This paper uses constructed knowledge thesaurus and the Snorkel method to weakly supervise the classification of unlabeled news with no manual tagging. Subsequently, we utilize LDA to generate the topics and obtain the signal value of each topic based on the topic signal detection function. Finally, we establish the temporal topic relationships and get the visualized results of news aggregation. One news dataset from one channel of Sina.com is used to test the framework. We obtain both the classification and aggregation of news, and generate the visualized results of the topics with topic signal and temporal evolution relationships.

Keywords: Weakly-supervised · News aggregation · Topic signal · Topic evolution

1 Introduction

People use a lot of information when making decisions. In the era of big data, how to quickly obtain concerned information from online media is very important. Using automated tools to aggregate information has become the focus of researchers.

When we process a large amount of news data, one considerable work is classifying the news. Many researchers use machine learning or deep learning methods and have obtained excellent results on text classification tasks [1–4]. These methods rely on a large number of labeled training sets [5]. When lacking labeled data, the Snorkel method proposed by Stanford AI Lab works well. Ratner et al. [6,7]

J. Chen et al. (Eds.): KSS 2022, CCIS 1592, pp. 50–64, 2022.
https://doi.org/10.1007/978-981-19-3610-4_4

presented a weakly supervised classification architecture that requires only a tiny amount of human-provided information. In our work, we use the Snorkel method to complete the multi-classification problem of unlabeled datasets.

In addition to the classification of news, the news topics are also of consideration. Blei et al. [8,9] proposed the famous Latent Dirichlet Allocation, a generative probabilistic model for collections of discrete data such as text corpora. The method combines the bag-of-words model to identify documents' topics and topic content from many texts in an unsupervised machine learning manner. As for the evolutionary relationships of topics, the Sankey diagram is a helpful visualization tool. Yan and Tang[10] used the LDA and the Kleinberg model to extract and filter the emergent topics of the "safety helmet event." They used Jaccard similarity to measure the evolutionary relationship among topics over time series and constructed a thematic evolution Sankey diagram of "safety helmet event." Similarly, Pépin et al. [11]discussed the influence of different metric distances on the calculation of topic relations, considered the weight of topic between consecutive time slots, and used Sankey diagrams to analyze topic evolution and topic signal with Twitter data.

Although combining LDA with the Sankey diagram helps capture the topic evolution effectively, the steadiness and popularity of the topics are also noteworthy. Thorleuchter and Van [12] proposed a method for weak signal identification using latent semantic indexing to detect weak signals in Twitter data and helped strategic planners to react ahead of time. Akrouchi et al. [13] proposed a fully automatic LDA-based weak signal detection method, which combined multiple indicators to define a signal detection function to find potential weak signal topics in a topic set. In our study, we further improve these methods for identifying topic signals.

In this paper, as shown in Fig. 1, we present an end-to-end weakly-supervised news aggregation framework to assist expert user in tracking the evolution of important content and topics from news. Our contributions are as follows:

- Provide an end-to-end framework whose input is unannotated text information and output is fully processed and visualized results.
- Introduce a weakly supervised method to classify texts quickly.
- Combine several effective indicators to detect the signal value of each topic.
- Measure the adjacent period's topic relationship based on the temporal topic signal value.

We test the feasibility and validity of our framework on a dataset covering the data from the military board of Sina.com in 2016. The result shows that our framework effectively aggregates news. Moreover, the signal detection algorithm helps find the temporal topics with strong signals and strong correlations.

2 Methodology

2.1 Classification Method Based on the Snorkel

In recent years, machine learning models have been extensively used in text classification tasks and have achieved excellent results. These models generally

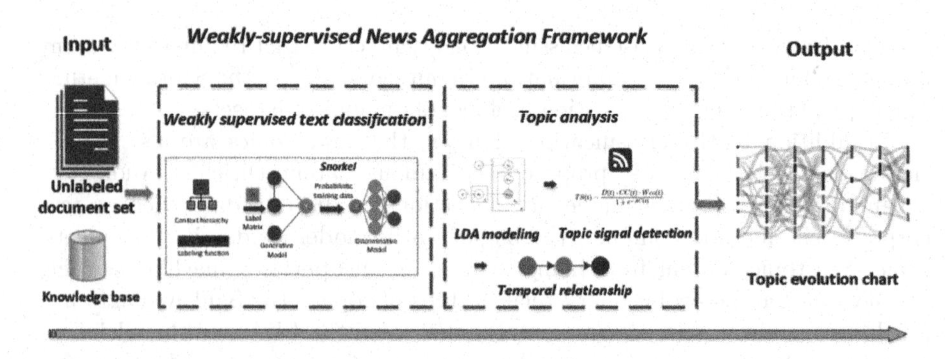

Fig. 1. End-to-end weakly-supervised news aggregation framework. The input of the framework is an unlabeled document set and knowledge base, and the output is a visual topic evolution Sankey diagram. The framework includes two parts: weakly supervised text classification and topic analysis. The topic analysis includes LDA modeling, topic signal detection, and temporal topic relationships detection.

rely on massive hand-labeled training datasets, which is expensive and time-consuming. As a result, acquiring well-labeled training data becomes a significant prerequisite and difficulty for machine learning. In this context, many weakly supervised learning methods [14–17] have been presented to obtain noisy but efficient data labels. This paper adopts the Snorkel weakly supervised machine learning architecture, allowing users to write multiple weakly supervised programs to rapidly generate, manage, and model training data. As shown in Fig. 2, the Snorkel method mainly has the following three stages:

1) **Writing the labeling functions**: On the premise of no hand-labeling data, Snorkel users write heuristic labeling functions to generate a label or abstain for the unlabeled sample. The labeling function allows the users to specify a wide range of weak supervision sources such as patterns, heuristics, external knowledge bases, and more;
2) **Modeling a generative model**: According to the labeled results generated by the labeling functions, the Snorkel automatically trains a generative model to catch the agreements and disagreements of the labeling functions. The generative model is used to estimate the accuracies of the different labeling functions and then to re-weight and combine their labels to produce a set of probabilistic training labels, effectively solving a novel data cleaning and integration problem. Several experiments results show that the accuracy of the final label has been significantly improved after such processing;
3) **Training a discriminative model**: Users train a discriminative model with high coverage and robustness based on the probabilistic labels from the generative model. As more data is provided, the discriminative model sees more features that co-occur with the heuristics encoded in the labeling functions, and the error will decrease. The standard discriminant models include machine learning models such as decision trees and random forests and deep learning models like CNN and LSTM.

Based on the three stages described above, both unlabeled data and weak supervision sources are the input into the Snorkel Pipeline, which ultimately outputs labeled data. After checking the output results, we modify the labeling functions to improve the method's performance through positive feedback.

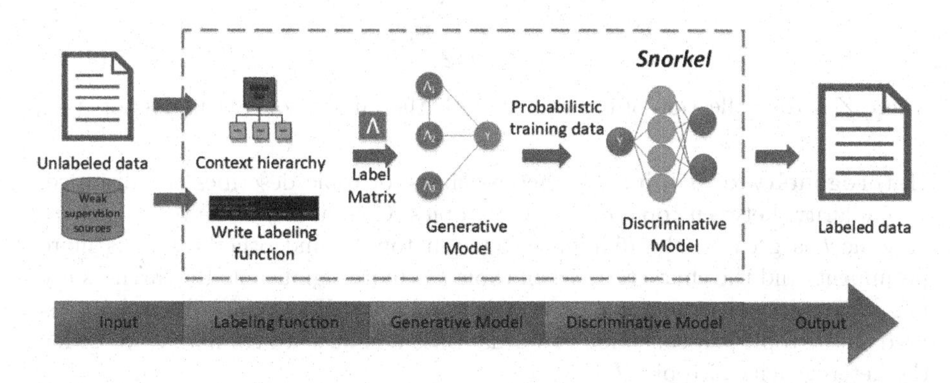

Fig. 2. An overview of the Snorkel architecture. Inputs are unlabeled data and weak supervision sources. The Snorkel architecture includes the labeling function, generative model, and discriminative model.

2.2 Topic Modeling and Topic Signal Detection

Topic models have received particular interest as text data have been increasing extensively. They are widely used to seek latent information from text corpora to capture some text features that are difficult to find manually. Therefore, the topic models play a critical role in our overall framework. The most conventional and effective topic model is the LDA, which assumes the topics and words in a document follow two Dirichlet distributions. We use the LDA algorithm tool in Python's Gensim library and determine the number of topics by coherence and perplexity.

The topics we get with LDA are usually various and complex, making it challenging to find the key topics and relevant news. Hence, the concept of the topic signal is adopted to reflect the hot degree, persistence, and stability of a topic. By the topic signal, the influential and noteworthy topics are quickly found. We use the topic signal to understand the main news content in this period. In comparison, a weak signal topic may contain some imperceptible news initially but may break out later and be a worthy concern.

Referring to previous approach [12,13], we define the signal detection function based on four primary indicators: focused degree of topic, heterogeneity of topic, coherence weight of topic, and autocorrelation coefficient of topic.

Focused Degree of Topic: The focused degree of topic represents the proportion of text related to the topic to the total number of text. This indicator directly reflects the signal strength of the topic in the current period time.

A topic's focused degree is high, indicating that this topic is highly concerned by the public and dramatically influences society. For topic t, the focused degree of topic $D(t)$ is expressed by dividing the number of text belonging to the topic by the total number of text within the time :

$$D(t) = \frac{N_t}{\sum_{t \in Z} N_t} \tag{1}$$

where Z is the collection of topics and N_t is the number of text under topic t.

Heterogeneity of Topic: The heterogeneity of topic describes the distance, or similarity, between topic t and other topics. Commonly, if the heterogeneity of topic t is greater, the difference between topic t and other topics is more prominent, and the characteristics of topic t is more significant. Researchers use many distance measures to measure the similarity. Here Jaccard similarity is used as the topic is a collection of words. Based on Jaccard similarity, we define the heterogeneity of topic $H(t)$:

$$H(t) = \frac{1}{\sum_{t_v \neq t} d(t, t_v)} = \frac{1}{\sum_{t_v \neq t} (1 - J(t, t_v))} \tag{2}$$

where t_v is the topic in the collection of topics and $J(t, t_v)$ is the Jaccard similarity between topic t and t_v.

Coherence Weight of Topic: The coherence weight reflects the internal properties of the topic t. This measure is based on the coherence of the relevant topic, which scores topic by measuring the semantic similarity within topic of words. The weight of a topic is the value assigned based on the significance of each topic. The calculation of topic coherence here is consistent with the LDA modeling process. Hence we define the coherence weight of topic Wco_t as follow:

$$Wco(t) = \frac{Coh(t)}{\sum_{t \in Z} Coh(t)} \tag{3}$$

where Z is the collection of topics, $Coh(t)$ is the coherence score of topic t.

Autocorrelation Coefficient of Topic: Considering the numbers of the text belonging to topic t can be regarded as a time sequence, as an indicator in time series analysis, the autocorrelation coefficient can be used to measure the lagged relationship of the numbers of texts in the topic t. The autocorrelation coefficient describes the correlation between the values of random signals at the time i and time $i + k$, which measures the influence of one's past behavior on one's present. This indicator effectively determines whether a topic remains relatively constant in its exposure over that time. Therefore, we define the autocorrelation coefficient of topic $AC(t)$ as follows:

$$AC(t) = \frac{Cov(t)_k}{Var(t)} \tag{4}$$

where $Cov(t)_k$ is the covariance of topic t at lag k and $Var(t)$ is the variance of topic t.

Topic Signal Detection Function: After defining the above four basic indicators, we establish the topic signal detection function TS based on the logistic function. Generally speaking, the logistic function is a mathematical model used to explain the law of population growth or biological population growth. Given the four functions defined above, (1), (2), (3) and (4) are used to form the topic signal detection function TS as follows:

$$TS(t) = \frac{D(t) \cdot H(t) \cdot Wco(t)}{1 + e^{-AC(t)}} \tag{5}$$

Previous studies use the topic signal function to discover the topics with weak signals by potential warning filtering functions [13]. Instead of focusing on weak signal detection, this paper focuses on locating those influential topics and finding related news with topic signals for news aggregation.

2.3 Temporal Topic Relationship Detection

A Sankey diagram can represent the temporal evolution of topics. We divide the data into multiple periods, use LDA to capture their topics, and build a Sankey diagram of topic evolution. Previous studies [10] utilize some text distance measures to judge the relationship between two topics, while only considering the textual similarity of topics and ignoring other critical indicators like the topic signal.

In this subsection, we deal with the problem of the relationship between temporal topics by combining the previous topic signal score. The Sankey diagram shows that the temporal topic relationship is considered to be modeled by a layer-weighted digraph G. The vertices represent the topics, and the directed edge denotes the relationship between the topics on adjacent layers in time. We consider that the news document set $\mathbb{D}$ is recorded during a time length T. Time length T is divided in n consecutive period $T_1, T_2, ...T_n$ and the corresponding news set is $D_1, D_2, ...D_n$. For each period T_i, we train a LDA model on the corresponding news set D_i for extracting timely focused topics set Z_i. Then we analyse relationship between topic $z \in Z_i$ and $z' \in Z_{i+1}$. Jaccard similarity is used to measure the similarity between two topics. In addition, topic signals are also important for temporal topic relationships. Generally, the topic of high signal value is likely to continue to appear in the next layer. Hence, we define the relationships between temporal topics as follows:

$$W(z, z') = (TS(z) + TS(z')) \times J(z, z') \tag{6}$$

where $TS(z)$ is the signal value of topic z, and $J(z, z')$ is the Jaccard similarity between topic $z \in Z_i$ and topic $z' \in Z_{i+1}$. What is noteworthy is that before using Jaccard similarity to calculate the temporal topic similarity, the top r words in probabilities are selected from the related topic word set to represent the topic.

3 Experiment and Result Analysis

Using automated tools for news aggregation and visualization is significant to help extract features of news data. News data from one channel of Sina.com in 2016 are selected to test the performance of the end-to-end weakly-supervised news aggregation framework mentioned in the previous section. The whole data contains 14,392 news items, including news headlines, main news body, and released time.

3.1 Weakly Supervised News Text Classification

We divide those news into five categories: military, politics, economics, technology, and other. The data are all unlabeled, and using conventional supervised classification methods will cost us a lot of time on data labeling. Thus, we use the Snorkel structure mentioned above to achieve weakly-supervised classification tasks for military, politics, economics, technology, and others five target categories.

As shown in Fig. 2, we construct the labeling functions for each category. Generally, the input of the labeling function is unlabeled data, and the output is a candidate class label or abstains label. The labeling function is mainly used to judge whether the input news contains words in the knowledge base corresponding to the category.

Based on the classification category, we construct the domain knowledge base from the internal knowledge base and the external knowledge base respectively. The internal knowledge base includes the experience thesaurus, and the external knowledge base includes common vocabulary thesaurus and external knowledge thesaurus. The construction process of thesaurus has the following three parts:

1) Internal knowledge base - experience thesaurus. We randomly extract several pieces of news data and judge the category according to the news content. Then we use the TextRank algorithm [18] to extract the top 20 keywords in the news headlines and text content and combine the keywords in the news to form an experience thesaurus.
2) External knowledge base - common vocabulary thesaurus. To effectively supplement the experience thesaurus, we collect the commonly used proprietary vocabulary corresponding to each category from Sogou Pinyin's official website[1], including four categories of the military, politics, economics, and technology.
3) External knowledge base - external knowledge thesaurus. To further expand the external knowledge base, according to the news data set contributed by the

[1] https://pinyin.sogou.com/dict/.

Fudan University team[2], three categories of military, politics, and economics are extracted, and the categories related to technology are combined as the category of technology. Then we use the Textrank algorithm to extract the keywords in the four categories and build the external knowledge thesaurus.

According to the domain knowledge base constructed above, the labeling functions of each category are respectively constructed, wherein each labeling function corresponds to a domain knowledge thesaurus. If the news text contains the word in the knowledge thesaurus, the news is considered to belong to the corresponding category, and the corresponding label is output. We build 13 labeling functions. The news in the dataset, the labeling function, and the corresponding domain knowledge base are input into the Snorkel architecture. We choose the decision tree as the discriminant model and get the categories of all news.

We extract 2000 news whose titles contain "Nanhai" for manual labeling to test our classification results. Comparing the experimental results and manual labeling results, the weighted accuracy of our training results is 66.66%, the weighted recall rate is 63.82%, and the weighted F1-score value is 64.77%. The result shows that the weakly-supervised classification method can still effectively identify news categories under the condition that it takes a shorter time. One reason that affects our classification performance is the imbalance of the sample data. Using this method, we obtain high-quality data while significantly reducing the cost of manual labeling through the iterative learning method of small sample learning - machine learning classification -submission of manual verification - correction of learning results.

3.2 Topic Modeling and Topic Signal Analysis

In this subsection, we utilize LDA to get topics and calculate the topic signal scores and temporal topic relationships. To find the optimal number of topics, we train multiple models for the different numbers of topics separately and use the topic coherence score to judge. In this experiment, after numerous pieces of training, we find that when the number of topics is 20, the topic coherence score reaches the highest value of 0.558. Since the time of the sample data is from January to December 2016, it is divided into twelve stages $T_1 ... T_{12}$ by month. With the preset number of topics is 20, we train the LDA model on the data of each stage separately. We choose the top 50 words with the highest probability value to represent the topic for each topic.

According to the Eq.(5) defined above, we calculate the signal value for each topic. We use the specified Eqs.(1)–(4) to calculate the focused degree of topic, heterogeneity of topic, coherence weight of topic, and autocorrelation coefficient of topic separately. When calculating the autocorrelation coefficient, we set the lag coefficient to 14 because of the general thinking that the duration of the hot topic after the emergence is 14 days [13]. Finally, we use the combination of these four indicators to get the signal value of the topic.

[2] https://www.heywhale.com/mw/dataset/5d3a9c86cf76a600360edd04.

Figure 3 shows some calculation results toward news in January. The autocorrelation coefficient of Topic 2 of the news in January is 0.039272, which indicates that Topic 2 has a weak lag correlation in time. And the autocorrelation coefficient of Topic 6 of the news in January is -0.371470, which shows that Topic 6 is negative self-correlation in time. For further explanation, Fig. 3(a) shows the number of daily news for Topic 2 and Topic 6 in January news. The vertical axis value in the figure is the difference between the number of news on that day and the monthly average. Topic 2's daily news numbers are evenly distributed on both sides of the coordinate axis, showing a relatively stable state along the time. However, Topic 6 has a reasonably noticeable lag in time. The number of news on this topic is at a high level on the 6th days and 7th days but decreases significantly after 14 days. Thus, the topic autocorrelation coefficient reflects the topic's stability and persistence in time.

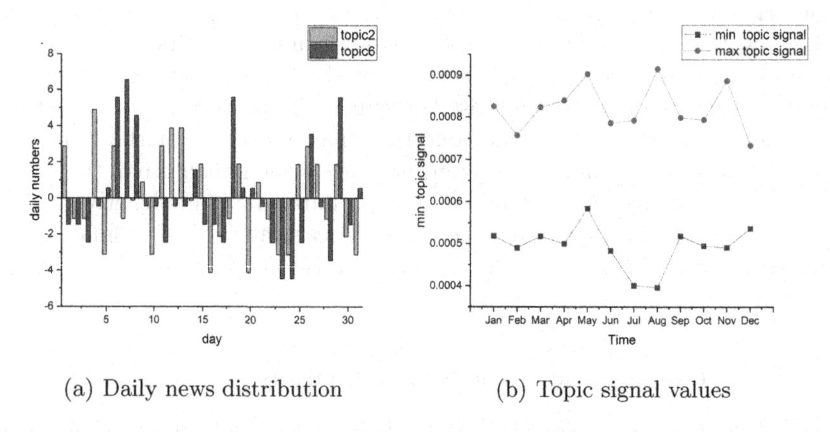

(a) Daily news distribution (b) Topic signal values

Fig. 3. Topic signal detection example in January. (a) The numbers distribution of daily news of Topic 2 and Topic 6 in January. The ordinate is the number of news the day in a topic minus the topic's daily average in January. (b) The maximum and the minimum topic signal values at each month.

Figure 3(b) shows the maximum and minimum topic signal values in each stage. We find a clear difference between the maximum and minimum values, and the former is almost twice as large as the latter. There is indeed a difference in the strength of the topic signal in the topics we obtain. Tabel 1 lists the keyword and news examples of three strong signal topics in January. Topic 1 is about the nuclear test that took place in North Korea. The main keywords of Topic 1 are North Korea, nuclear test, Park Geun-Hye, and nuclear weapons. Topic 5 is about China's military drills in the Northwest Desert, and Topic 10 is about Japan ordered the deployment of Aegis ships to intercept North Korean missiles. With the help of LDA and detection of the topic signal, we quickly locate strong signal topics that have attracted widespread attention during this period and weak signal topics and news with weak signal values but are likely to ferment in the future.

Table 1. Topics in January news

Topics#	Topic keywords	News example
Topic1	朝鲜 (North Korea), 核试验 (nuclear test), 朴槿惠 (Park Geun-hye),核武器 (nuclear weapon)	朝鲜核试验令日韩靠拢 ,或分享敏感军事情报 (North Korea's nuclear test arouse Japan and South Korea to share intelligence.)
Topic5	火箭军 (Rocket Force),战机 (warplane),解放军 (PLA),导弹 (missile)	解放军火箭军多支导弹旅进行实战训练 (PLA Rocket Force conducts training exercises.)
Topic10	部 署 (deploy),部 队 (troops), 回 应 (respond), 拦截 (intercept)	日本已下达摧毁令 ,部署宙斯盾舰拦截朝鲜洲际导弹 (Japan deploys Aegis ships to intercept North Korean ICBMs.)

3.3 Temporal Topic Relationship Detection

After locating the crucial topics and news using LDA and topic signal detection, we next explore how these topics have evolved and the relationship among the topics over time. We utilize the function (6) to discover the relationship between temporal topics. As mentioned above, the temporal topic relationship can be constructed as a layer directed graph G, where the vertices represent topics, and the directed edges represent the evolutionary relationship from the previous topic to the next topic in time. The temporal topic relationship $W(z, z')$ of the directed edge depends on the signal values of the two topics and their similarity. Using the Jaccard similarity and the signal value of each topic, we calculate the relationship value between topics in two adjacent periods. We extract the three topics with the highest temporal relationship value and get Fig. 4.

The two topics on the left of Fig. 4 are Topic 2 and Topic 6 in the news of March, and on the right is Topic 1 in the news of April. The stacked bar on the right of the topic is the category distribution of news with the top 50 probability values in the topic. Mar.Topic 2 and Apr.Topic 1 are mostly military-related news, while Mar.Topic 6 contains more political news. News with categories of economics, technology, and others make up only a tiny portion. Relevant news shows the probability that the news belongs to this topic, the title of the news, and the category of the news. Two examples of news with the highest probability are given in each topic. These three topics are about disputes between China and surrounding countries on the maritime border. Mar.Topic 2 is the military operations of Japanese submarines in the East China Sea and the South China Sea. Mar.Topic 6 is the diplomatic position of China and the United States on military issues in the South China Sea. Both Mar.Topic 2 and Mar.Topic 6 point to Apr.Topic 1, which shows China's ongoing dispute with Japan and the USA in the South China Sea. These three topics are centered on the South

China Sea issue and are strongly related to content. This result shows that our framework effectively aggregates news according to the content and discovers temporal topics with strong signals and strong correlations.

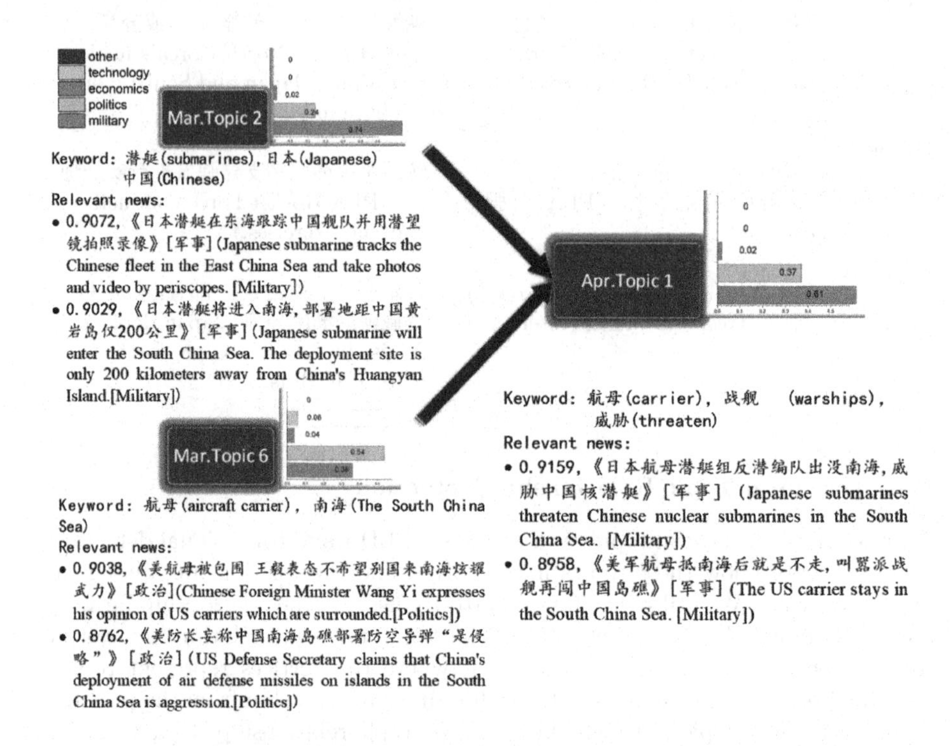

Fig. 4. Examples of topic evolution. The stacked bar on the right of the topic shows the proportion of news categories with the top 50 probability values in the topic. Keywords and relevant news are displayed under the topic. Relevant news shows the probability that the news belongs to this topic, the title of the news and the category of the news.

Table 2. Topic values

Topics#	Focused degree	TS values($\times 10^{-3}$)
Jan.Topic 1	98	0.65072
Jan.Topic 12	71	1.00060
Feb.Topic 4	88	0.77016
Feb.Topic 15	63	1.01864
Mar.Topic 2	88	0.87274
Mar.Topic 17	82	0.92985
Apr.Topic 7	70	0.87855

To illustrate the effectiveness of topic signal detection and temporal relationship in our framework, we use previous topic evolution method [10]for comparison. Previous method does not discuss the topic signals and only uses the Jaccard similarity to measure the temporal relationship of topics. Figure 5 shows two main topic evolution chains identified by previous method and the method of our framework. Figure 5 shows part of the Sankey diagram based on the Sina military data set, which shows the topics and the topic evolution relationship generated from January to April. The blue topic in Fig. 5 is the largest signal topic found based on the topic signal detection method in our framework, while the red topic is the largest focused degree topic in previous method. We find that the results of the other three months are different, except that the Apr. Topic 7 is considered by both methods to be the most influential topic. In addition, it can be seen from Table 2 that Jan.Topic 1, Feb.Topic 4, and Mar.Topic 2 are the topics of the largest focused degrees in that month, but their signal values are not the strongest. Since topic signals and Jaccard similarity determine the temporal relationship between topics, the topic with a larger signal value is more likely to be considered as part of the main topic evolution path.

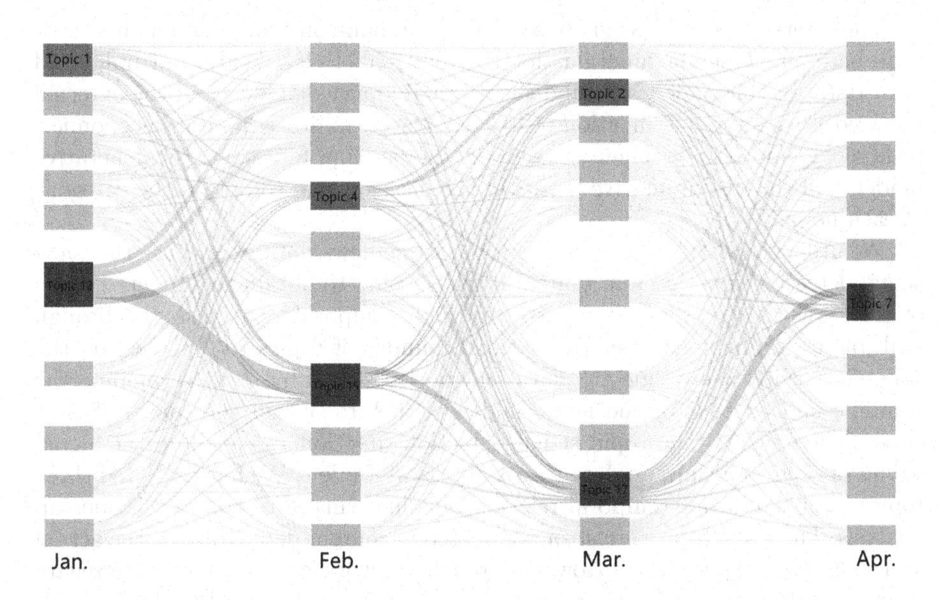

Fig. 5. Compare the maximum impact topics detected by different methods. The red topic is the result of the comparison method, and the blue is the result of our framework.

In terms of content, the blue topics from January to April are mainly about China's military activities and related diplomatic actions in the South China Sea with neighboring countries. The topic evolution chain formed by blue topics has remarkable correlation and consistency in the temporal relationship. In contrast, the red topic evolution chain is about North Korea's nuclear test and the dispute between China and Japan over the Diaoyu Dao. But there is no significant correlation in the red topics. Even if previous method can identify the red topic evolution chain by capturing the evolving relationship between topics with a highly focused degree, these topics are not significantly related in content. In contrast, the topics in the blue topic evolution chain identified by the topic-signal-based method proposed in this paper have strong signals and have consistent or similar topic content. Our framework has resulted in a more effective solution of digging out the topic evolution relationship through the above comparison.

4 Discussion and Conclusions

This paper presents a novel end-to-end weakly supervised news aggregation framework. We combine the Snorkel method, Latent Dirichlet allocation, topic signal detection, and temporal topic relationship detection to aggregate unlabeled news and obtain effective visualization results. We use a dataset covering the data from the military bound of Sina.com in 2016 to test the performance of our framework.

The Snorkel architecture is based on the human-constructed domain knowledge base and labeling functions for the multi-label classification of unlabeled news data. In this paper, we build the internal knowledge base and the external knowledge base for classification of five categories: military, politics, economics, technology, and others. Then we combine the labeling functions, the generative model, and the discriminative model of the decision tree to obtain the category of each news.

We utilize the LDA to obtain the topic of the news dataset and use the topic signal detection function and temporal relationship to obtain the directed edges of topics with high signal strength and high temporal relationship. Through multiple experiments, we set the optimal number of topics to 20 based on the coherence score. We divide the data into twelve stages by month, combine the headline and context of the news, and use LDA to obtain the topics of each stage. Then we use the topic signal detection function to calculate the signal values of each topic and analyze the examples in Table 1. We have selected three topics with significant temporal relationships in terms of temporal relationship analysis. The three topics are strongly related in content depending on the actual situation. Our experiments show that our framework can effectively aggregate news according to news content without manual annotation. In addition, the signal detection algorithm we use also allows us to discover temporal topics with strong signals and strong correlations.

Our work aims to provide decision-makers with an efficient tool that automatically aggregates news while reducing time-consuming and saving costs. The experiment results indicate that our framework can efficiently aggregate news by topic, generate the signal value of topic, and provide a reliable reference for decision-makers. In the future, we will further improve the performance of our framework, such by adding the function of crucial event detection. In addition, the visualization of the results needs to be improved in the future.

Acknowledgements. This paper is supported by the National Natural Science Foundation of China (71731002 & 71971190).

References

1. Li, C., Zhan, G., Li, Z.: News text classification based on improved Bi-LSTM-CNN. In: 9th International Conference on Information Technology in Medicine and Education (ITME), pp. 890–893. IEEE (2018)
2. Miao, F., Zhang, P., Jin, L., et al.: Chinese news text classification based on machine learning algorithm. In: 2018 10th International Conference on Intelligent Human-Machine Systems and Cybernetics (IHMSC), vol. 2, pp. 48–51. IEEE (2018)
3. Wu, M.J., Fu, T.Y., Chang, Y.C., et al.: A study on natural language processing classified news. In: 2020 Indo-Taiwan 2nd International Conference on Computing, Analytics and Networks (Indo-Taiwan ICAN), pp. 244–247. IEEE (2020)
4. Minaee, S., Kalchbrenner, N., Cambria, E., et al.: Deep learning-based text classification: a comprehensive review. ACM Comput. Surv. (CSUR) **54**(3), 1–40 (2021)
5. Sun, C., Shrivastava, A., Singh, S., Gupta, A.: Revisiting unreasonable effectiveness of data in deep learning era. arXiv preprint arXiv: 1707.02968 (2017)
6. Ratner, A.J., Bach, S.H., Ehrenberg, H.R., et al.: Snorkel: fast training set generation for information extraction. Proceedings of the 2017 ACM International Conference on Management of Data, pp. 1683–1686 (2017)
7. Ratner, A., Bach, S.H., Ehrenberg, H., et al.: Snorkel: rapid training data creation with weak supervision. In: Proceedings of the VLDB Endowment. International Conference on Very Large Data Bases. NIH Public Access, vol. 11, no. 3, p. 269 (2017)
8. Blei, D.M., Ng, A.Y., Jordan, M.I.: Latent dirichlet allocation. J. Mach. Learn. Res. **3**(Jan), 993–1022 (2003)
9. Blei, D.M.: Probabilistic topic models. Commun. ACM **55**(4), 77–84 (2012)
10. Yan, Z.H., Tang, X.J.: Exploring evolution of public opinions on Tianya club using dynamic topic models. J. Syst. Sci. Inf. **8**(4), 309–324 (2020)
11. Pépin, L., Kuntz, P., Blanchard, J., et al.: Visual analytics for exploring topic long-term evolution and detecting weak signals in company targeted tweets. Comput. Ind. Eng. **112**, 450–458 (2017)
12. Thorleuchter, D., Van den Poel, D.: Weak signal identification with semantic web mining. Expert Syst. Appl. **40**(12), 4978–4985 (2013)
13. El Akrouchi, M., Benbrahim, H., Kassou, I.: End-to-end LDA-based automatic weak signal detection in web news. Knowl.-Based Syst. **212**, 106650 (2021)
14. Zhou, Z.H.: A brief introduction to weakly supervised learning. Natl. Sci. Rev. **5**(1), 44–53 (2018)
15. Mahajan, D., Girshick, R., Ramanathan, V., et al.: Exploring the limits of weakly supervised pretraining. In: Proceedings of the European conference on computer vision (ECCV), pp. 181–196 (2018)
16. Medlock, B., Briscoe, T.: Weakly supervised learning for hedge classification in scientific literature. In: ACL, pp. 992–999 (2007)

17. Türker, R., Zhang, L., Alam, M., Sack, H.: Weakly supervised short text categorization using world knowledge. In: Pan, J.Z., et al. (eds.) ISWC 2020. LNCS, vol. 12506, pp. 584–600. Springer, Cham (2020). https://doi.org/10.1007/978-3-030-62419-4_33
18. Mihalcea, R., Tarau, P.: Textrank: bringing order into text. In: Proceedings of the 2004 Conference on Empirical Methods in Natural Language Processing, pp. 404–411 (2004)

Exploring the Asymmetric Effects of Perceived Quality on Product Evaluation: A Study of Automobile Review

Tong Yang(✉), Yanzhong Dang(✉), and Jiangning Wu

Dalian University of Technology, Dalian, China
yangt014@mail.dlut.edu.cn, yzhdang@dlut.edu.cn

Abstract. Perceived quality reflects consumers' perceptions of product attributes and then directly affects the overall evaluation. Existing research has concluded that the effect of perceived quality of attributes on overall evaluation is not symmetric, but research on this asymmetric effect within the automobile industry is uncommon. In this paper, we investigated the asymmetric effects of perceived quality on overall evaluation using consumer-generated data from Autohome. First, the effects of perceived quality on overall evaluation in different contexts was analyzed using penalty-reward contrast analysis (PRCA), and attribute classification was achieved by calculating the IA index, i.e., Appearance belongs to excitement attribute and the rest attributes belong to basic attributes. Second, the differences in the impact of different attribute categories were analyzed, and it was found that basic attributes were more important to overall evaluation in general. The results were obtained by robust standard errors-based OLS and Tobit, and were tested for robustness with data from different time periods. The related findings can help manufacturers prioritize attribute improvements, with basic attributes requiring more resources overall, but the attribute improvement priorities can be adjusted for different contexts.

Keywords: Perceived quality · Overall evaluation · Asymmetric effects · OLS · Automobile review

1 Introduction

Quality includes not only the objective quality of a product, but also the perceived quality generated by consumers in the course of use. This perceived quality is a subjective perception that varies from person to person and comes from the consumer's experience of using the product, which directly influences their evaluation of the product [1, 2] and further influences business behavior such as profitability, pricing, and market trust of firms [3, 4]. In the automobile industry, consumer overall evaluations are important to manufacturers due to the high cost of trial and error. Therefore, although there is no standard definition of terms related to perceived quality, the concept is already widely used by manufacturers including Nissan, Honda, Toyota, Mercedes-Benz, BMW, Audi, GM, Ford, Cadillac and others.

J. Chen et al. (Eds.): KSS 2022, CCIS 1592, pp. 65–80, 2022.
https://doi.org/10.1007/978-981-19-3610-4_5

In order to better study consumer overall evaluation, most researchers argue that product evaluation should be measured through multi-attribute performance [5–7]. Many previous studies have abstracted the relationship between perceived quality (attribute performance) and overall evaluation as linear or symmetric [5, 7–9], i.e., an increase or decrease in perceived quality by the same unit has the same degree of impact on the overall evaluation. However, some existing studies confirm that the relationship between perceived quality and overall evaluation is non-linear or asymmetric. That is, equal increases or decreases in perceived quality can have different effects on the overall evaluation [7, 10]. Research on the asymmetric impact of perceived quality can help manufacturers of automobiles identify categories of attributes and thus prioritize attributes in different contexts when improvements in overall evaluation are needed. However, few current consumer studies in the automobile field have focused on the asymmetric relationship between perceived quality and overall evaluation. Our research question is: What is the asymmetric effect of consumer perceived quality of different attributes of the automobile on the overall evaluation? How to determine the priority of attribute improvement according to attribute category?

This study investigated the asymmetric effects of consumer perceived quality on overall evaluation using social media review data from the automobile industry. First, the asymmetric effect of perceived quality was confirmed using the penalty-reward contrast analysis (PRCA) [11], which classifies attributes into basic and excitement attributes, and the classification results were validated using OLS and Tobit models. Second, the robustness of the results was demonstrated by validating data from different years to obtain the same results as above. Classification of vehicle attributes considering asymmetric effects can help manufacturers to prioritize attributes in different situations where there are differences in perceived quality. However, it is often difficult for manufacturers to change the resource allocation to different attributes in a short period of time, so it is necessary to determine the overall priority of basic and excitement attributes in order to optimize resources. Next, OLS was used to further investigate the specific relationship between the two types of attributes and the overall evaluation, and it was found that the perceived quality of the basic attribute had a greater positive effect on the overall evaluation than the excitement attribute, while both objective quality and expectation had a positive effect on the overall evaluation.

The study is structured as follows: Sect. 2 reviews the related works; Sect. 3 provides the research framework of the asymmetric effects of perceived quality using automobile social media data; Sect. 4 summarizes the findings; and after presenting the implications of the study, the final section concludes with a summary of the study's limitations.

2 Related Works

2.1 Effect of Perceived Quality on Overall Evaluation

Scholars have defined perceived quality in various ways, including subjective quality judgments relative to expected quality [12], all attributes in a product that exceed consumer expectations [13], and the extent to which a product fulfils its function [14]. The product evaluation process reflects how consumers instinctively learn and conceptualize the perceived quality of attributes and how connections are made between local concepts and overall evaluations. Based on the part-whole model of learning theory [15], overall evaluations are generated based on consumers' perceptions of each attribute of a product and the content descriptions associated with them. In social media contexts, consumers post content that contains many descriptions of product perceptions, which has a strong relationship with the overall evaluation of product.

It has been discussed that some components of consumption perceptions have a greater impact on overall evaluation than others. For example, in the field of food and beverages, Bai et al. [16] analyzed the different relationships between the perceived quality of food or service and the overall evaluation. In the field of information systems, McKinney, Yoon and Zahedi [17] analyzed information perceived quality and system perceived quality to obtain their different influence relationships on overall evaluation. In the field of re-engineered products, Abbey et al. [18] found that cosmetic attribute perceptions had a greater impact on overall ratings compared to functionality attribute perceptions. Therefore, the perceived quality of product attributes differs from category to category in terms of their impact on the overall evaluation.

2.2 Symmetric Versus Asymmetric Effects of Perceived Quality

Traditionally, studies of overall evaluations have used the expectation inconsistency paradigm [19], where the overall evaluation is the difference between expectations and perceived quality. Perceived quality that is lower than expectation is below the reference point, leading to dissatisfaction (negative variance). Perceived quality can also reach (be consistent) or even exceed the reference point (positive difference), both leading to satisfaction. A drawback of the expected inconsistency paradigm is that the effect size of these perceived quality deviations is assumed to be the same, although when the reference point is exceeded or is fallen below, it affects the overall evaluation of the product in the opposite direction. As a result, a symmetrical effect of perceived quality on the overall evaluation can naturally be derived. This view is challenged by the model of Kano et al. [20], which identifies three attributes of the quality category: must-be, performance and attractive attributes. The presence of the first category of attributes has only a minimal effect on overall evaluation enhancement, but absence or poor performance can have a strong negative effect on overall consumer evaluation. These attributes can be considered to have a negative asymmetry [21]. That is, adjusting the perceived quality level of a unit down for the same attribute has a greater impact on the overall evaluation than adjusting up. The second type of attribute describes the symmetric effect of the traditional perspective on overall evaluation: a downward or upward adjustment in the perceived quality level of a unit has the same effect on the overall evaluation. A high level of overall evaluation can be generated when the perceived quality of attractive attributes is increased, yet the absence of these attributes does not lead to dissatisfaction. Thus, they have a positive asymmetry, with an increase in the perceived quality of a unit having a greater effect on satisfaction than a decrease in a unit on dissatisfaction.

The theoretical logic for the existence of negative asymmetries lies in prospect theory, which means that people judge new options with a certain reference relevance and loss aversion [22]. Gains and losses arise from a comparison of reference points; above the reference point is considered a gain and below the reference point is considered a loss.

Loss aversion implies that a unit of loss is more important than an equivalent gain. Given that overall evaluations are also reference-related [23], prospect theory suggests that a one-unit decrease in perceived quality has a greater impact on overall evaluations than an increase. Positive asymmetry is rooted in consumer delight theory [19]. Consumer delight is "a very positive emotional state, usually resulting from exceeding someone's expectations by a surprising degree" [24]. The perceived quality of these attributes is high enough to make consumers happy, and they do not have a downward trend but an infinitely increasing trend.

The perceived quality of product attributes has a variable relationship to the overall product evaluation by consumers, i.e., an asymmetric effect. This phenomenon has been explored in several studies, which have constructed different names for attribute categories: satisfiers and dissatisfiers attributes [25, 26]; attractive and must-be attributes [20, 27]; or excitement and basic attributes [28, 29]. Attributes that can be categorized as excitement (attractive, satisfiers) have a higher impact on the overall evaluation at high levels of perceived quality than at low levels of perceived quality. Conversely, those attributes that can be classified as basic (must-be, dissatisfiers) have a higher degree of influence on the overall evaluation at low levels of perceived quality than at high levels of perceived quality.

3 Research Framework

Basic attributes are attributes that the product must have and do well or consumers will be extremely dissatisfied, whereas excitement attributes incur little dissatisfaction due to poor performance. Therefore, the priority of improvement of an attribute can be judged according to its category. Based on the above analysis, the research framework is proposed, the details of which are shown in Fig. 1.

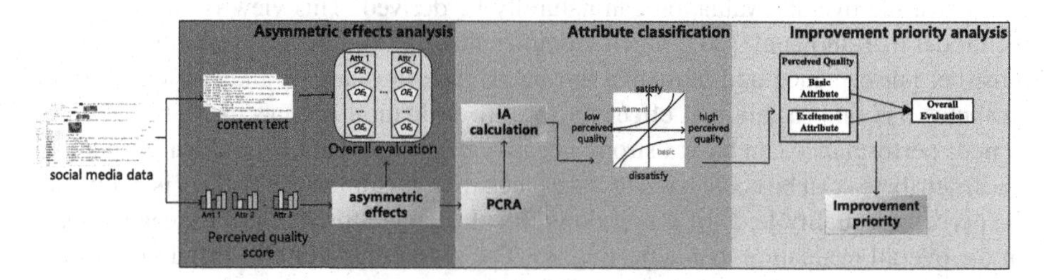

Fig. 1. Research framework

In studies exploring the asymmetry relationship between perceived quality and overall evaluation, the penalty-reward contrast analysis (PRCA) proposed by Brandt [11] has been widely used in the classification of attributes [30, 31]. Therefore, we also used this approach for attribute classification. PRCA enables the reconstitution of perceived quality scores for attributes by creating two dummy variables. d_{lp}^{i} is the first dummy variable that recodes the lowest perceived quality score (i.e., 1) as "1" and recodes the other scores (i.e., 2, 3, 4, 5) as "0", as a way of estimating the effect of lower perceived quality of the attribute on the overall evaluation [32]. d_{hp}^{i} is the second dummy variable that recodes the highest perceived quality scores (i.e., 5) as "1" and the other perceived quality scores (i.e., 1, 2, 3, 4) are recoded as "0", as a way of estimating the effect of higher perceived quality of the attribute on the overall evaluation [32]. Based on these two variables, we can use a multiple linear regression equation as in Eq. (1), where β_0, β^i_{lp} and β^i_{hp} are its coefficients.

$$OE = \beta_0 + \sum_{i=1}^{n} (\beta^i_{lp} d^i_{lp} + \beta^i_{hp} d^i_{hp}) + \varepsilon \tag{1}$$

After obtaining d^i_{lp} and d^i_{hp}, we used the *IA* index approach for attribute classification [30, 32], which is calculated as in Eq. (2).

$$IA_i = \frac{\left|\beta^i_{hp}\right| - \left|\beta^i_{lp}\right|}{\left|\beta^i_{hp}\right| + \left|\beta^i_{lp}\right|} \tag{2}$$

The *IA* index ranged from -1 to 1, with higher values indicating that the perceived quality of the attribute contributed more to the positive overall evaluation and lower values indicating that the perceived quality of the attribute contributed more to the negative overall evaluation. Following Albayrak and Caber [30] and Mikulic and Prebezac [32], we also set the classification threshold to 0.1. When $-1 \le IA_i < -0.1$, the attribute is a basic attribute; when $0.1 < IA_i \le 1$, the attribute is an excitement attribute; and when $-0.1 \le IA_i \le 0.1$, the attribute is not asymmetric, i.e., it is a performance attribute. After completing the classification of the attributes, we can obtain the average perceived quality of the basic and excitement attributes, denoted as PQ_{basic} and $PQ_{excitement}$. We continue to use Eq. (3) to analyze the relationship between perceived quality of different attribute categories and overall evaluation, where β_0, β_1, β_2 are its coefficients and X is a set of control variables.

$$OE = \beta_0 + \beta_1 PQ_{basic} + \beta_2 PQ_{excitement} + \beta_3 X + \varepsilon \tag{3}$$

4 Results

4.1 Measurement and Data

Social media provides a platform for consumers to freely express their feelings, and because of its comprehensiveness and timeliness, manufacturers and academics have taken a great interest in social media data. Autohome is the largest content exchange platform for automobile consumers in China, with 45 million daily users on mobile as of the first half of 2021. In this paper, we choose data from 2015–2020 on this platform for our empirical study, and the data example is shown in Fig. 2.

As in Fig. 2., the dataset contains 8 attributes, namely, Space, Power, Handling, Energy consumption, Comfort, Appearance, Interior, Cost performance. The perceived quality of each attribute is valued from 1 to 5. The dataset also contains the textual content of consumers' overall evaluation of the vehicle, including "most satisfying point", "most

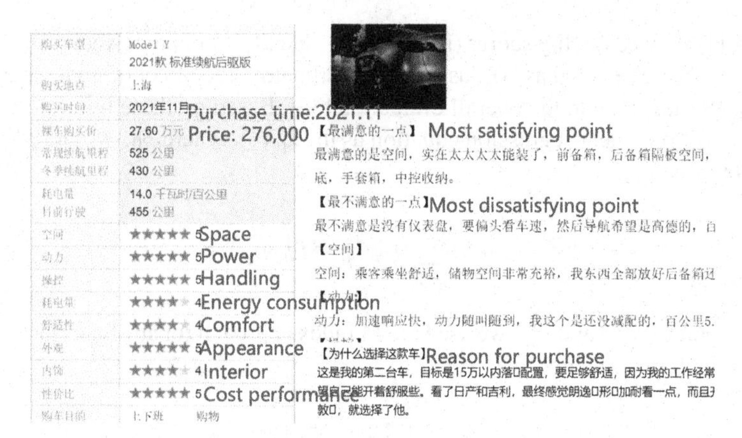

Fig. 2. Data example

dissatisfying point". We use the opinion extraction module of the AipNLP[1], which is an open-source natural language processing interface provided by Baidu, to calculate the overall evaluation score of the text content. The dataset also contains information on the purchase time, price, expectation. In addition, the number of complaints for each vehicle model was additionally crawled on a national car complaint website[2], to represent the objective quality of the car. These variables are added as control variables. Details of the specific variables are shown in Table 1.

Table 1. Summary statistics of key variables

Variables	Definition	Obs	Mean	Std. Dev.	Min	Max
OE	Score of consumer overall evaluation from 0 to 1	133355	0.741806	0.3619534	0.0005	0.9999
PQ_{space}	Score of consumer perceived quality of space from 0 to 1	133355	0.8844644	0.1757565	0	1
PQ_{power}	Score of consumer perceived quality of power from 0 to 1	133355	0.8565933	0.188017	0	1
$PQ_{handling}$	Score of consumer perceived quality of handling from 0 to 1	133355	0.8916726	0.1664911	0	1

(*continued*)

[1] https://ai.baidu.com/ai-doc/NLP/tk6z52b9z.

[2] http://www.12365auto.com/ranking/index.aspx?stime=2016-11-01&etime=2021-11-01&ba=0&bt=-1&cp=0&z=0&f=0&t=4&mip=2&mxp=110&ny=0.

Table 1. (*continued*)

Variables	Definition	Obs	Mean	Std. Dev.	Min	Max
PQ_{energy}	Score of consumer perceived quality of energy consumption from 0 to 1	133355	0.8688444	0.1973809	0	1
$PQ_{comfort}$	Score of consumer perceived quality of comfort from 0 to 1	133355	0.8055341	0.2059209	0	1
$PQ_{appearance}$	Score of consumer perceived quality of appearance from 0 to 1	133355	0.9284654	0.1410268	0	1
$PQ_{interior}$	Score of consumer perceived quality of interiors from 0 to 1	133355	0.8183514	0.2025047	0	1
PQ_{cost}	Score of consumer perceived quality of cost performance from 0 to 1	133355	0.8950339	0.1800412	0	1
OQ	Normalization of the number of complaints about one vehicle from 2016.11 to 2021.11	133355	0.1612015	0.2192926	0	1
expectation	Normalization of the number of attributes consumers expect before purchase	133355	0.444884	0.2884539	0	1
speed	Normalization of days between purchase and posting	133355	0.0210529	0.0422283	0	1
price	Normalization of vehicle prices	133355	0.0075039	0.0201915	0	1

After White's test, as in Table 2, the data were found to have significant heteroskedasticity, so in the subsequent regressions, OLS methods based on robust standard error were used.

Table 2. The results of White's test

Source	χ^2	df	p
Heteroskedasticity	6691.96	128	0.0000
Skewness	45862.75	16	0.0000
Kurtosis	279.05	1	0.0000
Total	52833.76	145	0.0000

4.2 Classification of Attributes Under Asymmetric Effects

First, we analyzed the asymmetry of the attributes according to the regression equation in Eq. (1). Also, we used a Tobit model to analyze the asymmetry of the attributes in the dataset to increase the confidence of the results (Table 3).

Table 3. Regression results for asymmetric relationships

Attribute	OLS ($R^2 = 0.1078$)			Tobit	
	Coef.		VIF	Coef.	
Space	β_{lp}^1	−0.04316** (0.019759)	1.06	β_{lp}^1	−0.04316*** (0.016376)
	β_{hp}^1	0.024799*** (0.002124)	1.13	β_{hp}^1	0.024802*** (0.002075)
Power	β_{lp}^2	−0.0966*** (0.018554)	1.1	β_{lp}^2	−0.09661*** (0.015786)
	β_{hp}^2	0.06009*** (0.00216)	1.25	β_{hp}^2	0.060083*** (0.002114)
Handling	β_{lp}^3	−0.09427*** (0.020987)	1.1	β_{lp}^3	−0.09427*** (0.017852)
	β_{hp}^3	0.056403*** (0.002353)	1.3	β_{hp}^3	0.056402*** (0.00224)
Energy consumption	β_{lp}^4	−0.07057*** (0.012744)	1.08	β_{lp}^4	−0.07058*** (0.010988)
	β_{hp}^4	−0.00448*** (0.002173)	1.2	β_{hp}^4	−0.00449** (0.002117)
Comfort	β_{lp}^5	−0.14551*** (0.014022)	1.11	β_{lp}^5	−0.14551*** (0.011991)
	β_{hp}^5	0.067133*** (0.001998)	1.23	β_{hp}^5	0.067138*** (0.0021)
Appearance	β_{lp}^6	−0.02629 (0.029317)	1.05	β_{lp}^6	−0.02629 (0.023665)

(continued)

Table 3. (*continued*)

Attribute	OLS ($R^2 = 0.1078$)			Tobit	
	Coef.		VIF	Coef.	
	β^6_{hp}	0.07741*** (0.002681)	1.25	β^6_{hp}	0.077412*** (0.002455)
Interior	β^7_{lp}	−0.10593*** (0.014103)	1.09	β^7_{lp}	−0.10593*** (0.01172)
	β^7_{hp}	0.033803*** (0.001984)	1.2	β^7_{hp}	0.033807*** (0.00206)
Cost performance	β^8_{lp}	−0.13446*** (0.014591)	1.12	β^8_{lp}	−0.13447*** (0.01246)
	β^8_{hp}	0.06247*** (0.002515)	1.34	β^8_{hp}	0.062467*** (0.00234)
constant		0.515848*** (0.002814)			0.515857*** (0.002451)

Notes. Standard errors (robust standard errors in OLS) in parentheses.
*$p < 0.1$ **$p < 0.05$ ***$p < 0.01$

After obtaining d^i_{lp} and d^i_{hp}, we used the *IA* index approach for attribute classification [30, 32]. According to the rules, the classification results of the attributes can be obtained as shown in Table 4.

Table 4. Results of attributes classification

Attribute	OLS		Tobit	
	IA	Category	IA	Category
Space	−0.27005	Basic	−0.27009	Basic
Power	−0.23299	Basic	−0.23309	Basic
Handling	−0.25128	Basic	−0.25132	Basic
Energy consumption	−0.8805	Basic	−0.88036	Basic
Comfort	−0.36854	Basic	−0.36856	Basic
Appearance	0.492945	Excitement	0.492969	Excitement
Interior	−0.51613	Basic	−0.51613	Basic
Cost performance	−0.36556	Basic	−0.36561	Basic

Table 4 shows the categories of vehicle attributes, with Appearance as an excitement attribute and the remaining attributes as basic attributes. According to the definition of the asymmetric effect, the perceived quality of most attributes of a vehicle has a strong influence on the negative overall evaluation of the consumer and is among the attributes that the manufacturer must get right. However, it has little impact on positive overall evaluations. Appearance, on the other hand, has a greater impact on the positive overall evaluation than on the negative overall evaluation. When perceived quality is low, consumers can live with it, but when perceived quality is high, it comes as a surprise to consumers.

The classification of attributes based on asymmetry is of great importance for manufacturers. The improvement of consumer perceived quality has always been a constant goal for manufacturers. However, manufacturers have limited resources and are often unable to work on multiple attributes at the same time when the perceived quality of all attributes needs to be improved. Therefore, the prioritization of attributes to be improved in different contexts is important to manufacturers. When the perceived quality of multiple attributes is poor at the same time, the basic attribute has a significant impact on the overall evaluation of the vehicle, so priority should be given to improving the basic attributes, i.e., Space, Power, etc. When the perceived quality of the attributes is high, the excitement attribute has a positive impact on the overall evaluation, the excitement attribute should be prioritized over the basic attribute, i.e., more resources should be invested in optimizing the Appearance.

4.3 Robustness

Consumer perceived quality is a subjective perception that is not static, therefore, to test the robustness of the above findings, it is necessary to analyze data from different years. According to data from the National Bureau of Statistics of China[3], the annual growth of vehicles rose before 2016 and fell after 2016, as shown in Fig. 3.

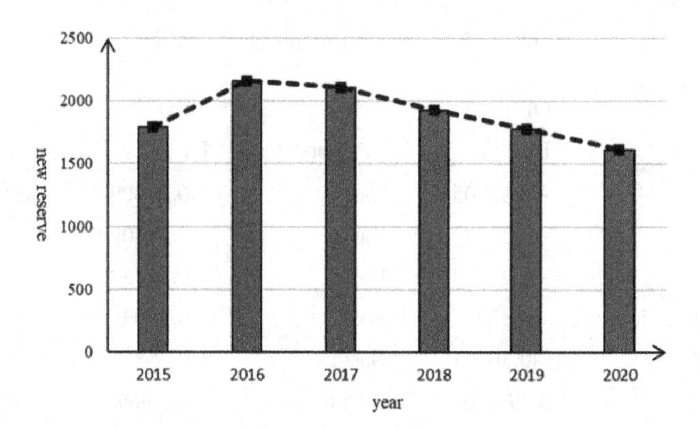

Fig. 3. Change in annual private passenger vehicle ownership in China (10,000 units)

Different vehicle purchase policies and social climates may affect perceived quality, making our conclusions change. Thus, we verified the findings of the attribute classification in two data sets, "2015–2016" and "2017–2020", and the results are shown in Tables 5 and 6.

[3] https://data.stats.gov.cn/easyquery.htm?cn=C01.

Table 5. OLS regression results for asymmetric relationship over time

Attribute	2015–2016			2017–2020		
	Coef.		VIF	Coef.		VIF
Space	β^1_{lp}	−0.0292298 (0.0269942)	1.06	β^1_{lp}	−0.068133** (0.0305925)	1.06
	β^1_{hp}	0.0150987*** (0.00445)	1.12	β^1_{hp}	0.0270955*** (0.0024911)	1.12
Power	β^2_{lp}	−0.0811167*** (0.0295551)	1.09	β^2_{lp}	−0.128221*** (0.0242709)	1.1
	β^2_{hp}	0.0251569*** (0.0046339)	1.19	β^2_{hp}	0.0571476*** (0.0025287)	1.2
Handling	β^3_{lp}	−0.0559345* (0.0328596)	1.09	β^3_{lp}	−0.124884*** (0.0281199)	1.1
	β^3_{hp}	0.044822*** (0.0047121)	1.26	β^3_{hp}	0.0540775*** (0.0028237)	1.26
Energy consumption	β^4_{lp}	−0.0564598*** (0.0219724)	1.08	β^4_{lp}	−0.0873937*** (0.0160905)	1.08
	β^4_{hp}	−0.0398888*** (0.0044249)	1.13	β^4_{hp}	0.0088239*** (0.0025729)	1.2
Comfort	β^5_{lp}	−0.123781*** (0.0219248)	1.11	β^5_{lp}	−0.1663789*** (0.0188052)	1.1
	β^5_{hp}	0.0480387*** (0.0051627)	1.32	β^5_{hp}	0.064706*** (0.0022105)	1.16
Appearance	β^6_{lp}	−0.0402126 (0.0415882)	1.06	β^6_{lp}	−0.0357591 (0.043025)	1.04
	β^6_{hp}	0.0641922*** (0.0049299)	1.26	β^6_{hp}	0.0711766*** (0.0034056)	1.18
Interior	β^7_{lp}	−0.0663525*** (0.0216228)	1.11	β^7_{lp}	−0.1405649*** (0.0190041)	1.07
	β^7_{hp}	0.0319276*** (0.0050075)	1.34	β^7_{hp}	0.0308211*** (0.0022046)	1.13
Cost performance	β^8_{lp}	−0.0998481*** (0.023762)	1.11	β^8_{lp}	−0.1528209*** (0.0194635)	1.13
	β^8_{hp}	0.0437091*** (0.0047056)	1.25	β^8_{hp}	0.0637042*** (0.0031204)	1.32
Constant		0.5221333*** (0.0046981)			0.538415*** (0.0039417)	
Observation		34104			90207	
R^2		0.0473			0.1065	

Notes. Robust standard errors in parentheses.
$*p < 0.1$ $**p < 0.05$ $***p < 0.01$

Table 6. Results of attributes classification over time

Attribute	2015–2016		2017–2020	
	IA	Category	IA	Category
Space	−0.31878	Basic	−0.43094	Basic
Power	−0.52656	Basic	−0.38342	Basic
Handling	−0.11029	Basic	−0.39565	Basic
Energy consumption	−0.17199	Basic	−0.81658	Basic
Comfort	−0.44082	Basic	−0.43998	Basic
Appearance	0.229679	Excitement	0.331204	Excitement
Interior	−0.35027	Basic	−0.64033	Basic
Cost performance	−0.39106	Basic	−0.41158	Basic

According to the results demonstrated in Tables 5 and 6, after dividing the dataset into two time periods and validating it in both OLS and Tobit models, the classification results for the attributes did not change, i.e., Appearance was the excitement attribute and the remaining attributes were the basic attributes. Thus, the robustness of the findings was confirmed within the context of the vehicle dataset of this study.

4.4 Effect of Perceived Quality of Different Attributes on Overall Evaluation

Attribute classification under asymmetric effects can help manufacturers identify attribute categories and thus prioritize perceived quality improvement when resources are limited. Since the context of perceived quality changes relatively frequently, the priority of vehicle attributes is in flux, but it is often difficult for manufacturers to switch their resource investment to different functional attributes within a short period of time, which leads to real-time adjustment of attribute perceived quality improvement priorities is not easy. Therefore, based on attribute classification under asymmetric effects, it is necessary to supplement the analysis of the overall cross-contextual priority order of different categories of attributes, and make the guidance to manufacturers more operable. We conducted an analysis of the effect of perceived quality of different categories of attributes on the overall evaluation according to Eq. (2), in which objective quality, expectations, price and the speed of consumer response were also considered as control variables, and the results are shown in Table 7.

The results presented in Table 7 show that the perceived quality of both the basic attribute and the excitement attribute significantly affect consumers' overall evaluation, and the basic attribute is more influential than the excitement attribute. Also, objective quality, speed and expectation significantly affect the overall evaluation.

From the above results, it can be found that, firstly, the perceived quality of basic attributes and excitement attributes both have a significantly higher degree of influence on the overall evaluation than other factors such as objective quality and expectation. This shows that consumers' overall evaluation of a vehicle mainly originates from the post-consumer experience segment, so manufacturers need to pay enough attention to

Table 7. Impact of perceived quality of different categories on overall evaluation

Variable	Coef.	VIF
PQ_{basic}	0.8068633*** (0.0098064)	1.38
$PQ_{excitement}$	0.1920607*** (0.0086361)	1.32
OQ	−0.0175099*** (0.0043552)	1.03
expectation	0.0850921*** (0.0035082)	1.08
price	−0.0904391 (0.0503443)	1
speed	−0.1323105*** (0.0259031)	1.07
Constant	−0.1620584*** (0.0088228)	
R^2	0.1169	

Notes. Robust standard errors in parentheses.
*p<0.1 **p<0.05 ***p<0.01

perceived quality. In addition, since the perceived quality of basic attributes has a greater impact on the overall evaluation than excitement attributes in general, manufacturers should incorporate this rule when prioritizing attribute improvement for different contexts based on attribute classification, i.e., keep more input for basic attributes in general and adjust the input of resources appropriately according to contextual changes.

5 Discussion and Conclusion

In this study, Autohome was selected as the data source to empirically analyze the asymmetric effect of attributes in the automobile industry. First, the strength of the effect of different attribute perceived quality on the overall evaluation positively and negatively was analyzed using the PRCA technique, and the classification of attributes was further achieved by calculating the IA index, where PRCA was implemented based on two models, robust standard error-based OLS and Tobit. Then, to test the robustness of the attribute classification results, experiments were conducted on data from different time periods, and the same results were obtained, i.e., Appearance as an excitement attribute and the rest of the attributes as basic attributes. The results of attribute classification can help manufacturers prioritize attribute improvement according to different levels of perceived quality. Next, we analyzed the effect of both types of attribute perceived quality on the overall evaluation. The results found that the degree of positive influence of the perceived quality of the basic attribute was stronger than that of the excitement attribute.

The theoretical implications of this study include, first, as one of the few studies examining the asymmetric effect of perceived quality on overall evaluation in the automobile industry, the asymmetric effect was confirmed and the vehicle attributes were classified. The relevant findings provide empirical arguments for subsequent studies on perceived quality in the automobile industry and vehicle improvement based on social media data. Second, based on the attribute classification under asymmetric effect, the strength of the impact of different categories of attribute perceived quality on the overall evaluation is analyzed, and basic attributes are found to have a stronger impact. It provides empirical evidence in the field of automobiles for asymmetric effect research, and provides some research ideas for future research. The practical implications are: first, the study helps manufacturers identify the categories of attributes and clarify the priority of attribute improvement in different contexts, thus improving efficiency when resources are limited; second, it obtains the overall priority of automobile attributes, which, combined with context-based prioritization, can help manufacturers maintain efficiency when resource switching is not timely.

Meanwhile, there are some limitations. First, the study data came from a single source, which is prone to certain sample selection biases. Future research could try the integration of multiple social media data and the combination of social media data and structured survey data. In addition, with the development of social media, there are various ways of product evaluation, such as pictures, voice, video, and long articles, but this study only focused on perceived quality scores and overall evaluation of short texts, which may leave out some information. It is hoped that multimodal data will be included in future studies.

References

1. Bolton, R.N., Lemon, K.N., Verhoef, P.C.: Expanding business-to-business customer relationships: modeling the customer's upgrade decision. J. Mark. **72**(1), 46–64 (2008)
2. Golder, P.N., Mitra, D., Moorman, C.: What is quality? An integrative framework of processes and states. J. Mark. **76**(4), 1–23 (2012)
3. Harutyunyan, M., Jiang, B.: The bright side of having an enemy. J. Mark. Res. **56**(4), 679–690 (2019)
4. Yang, Z., Sun, S., Lalwani, A.K., Janakiraman, N.: How does consumers' local or global identity influence price–perceived quality associations? The role of perceived quality variance. J. Mark. **83**(3), 145–162 (2019)
5. Chen, L.F.: Exploring asymmetric effects of attribute performance on customer satisfaction using association rule method. Int. J. Hosp. Manag. **47**, 54–64 (2015)
6. Mihalič, T.: Performance of environmental resources of a tourist destination: concept and application. J. Travel Res. **52**(5), 614–630 (2013)
7. Slevitch, L., Oh, H.: Asymmetric relationship between attribute performance and customer satisfaction: a new perspective. Int. J. Hosp. Manag. **29**(4), 559–569 (2010)
8. Chen, L.F.: A novel framework for customer-driven service strategies: a case study of a restaurant chain. Tour. Manage. **41**, 119–128 (2014)
9. Liu, Y., Bi, J.W., Fan, Z.P.: Ranking products through online reviews: a method based on sentiment analysis technique and intuitionistic fuzzy set theory. Inf. Fusion **36**, 149–161 (2017)

10. Bi, J.W., Liu, Y., Fan, Z.P., Zhang, J.: Exploring asymmetric effects of attribute performance on customer satisfaction in the hotel industry. Tour. Manage. **77**, 1–18 (2020)
11. Brandt, R.D.: A procedure for identifying value-enhancing service components using customer satisfaction survey data. Add Value Your Serv. **6**(1), 61–65 (1987)
12. Mitra, D., Golder, P.N.: How does objective quality affect perceived quality? Short-term effects, long-term effects, and asymmetries. Mark. Sci. **25**(3), 230–247 (2006)
13. Oxenfeldt, A.R.: Consumer knowledge: its measurement and extent. Rev. Econ. Stat. **32**, 300–314 (1950)
14. Box, J.M.F.: Product quality assessment by consumers — the role of product information. Ind. Manag. Data Syst. **83**(3/4), 25–31 (1983)
15. Swanson, R.A., Law, B.D.: Whole-part-whole learning model. Perform. Improv. Q. **6**(1), 43–53 (1993)
16. Bai, X., Marsden, J.R., Ross, W.T., Jr., Wang, G.: A note on the impact of daily deals on local retailers' online reputation: mediation effects of the consumer experience. Inf. Syst. Res. **31**(4), 1132–1143 (2020)
17. McKinney, V., Yoon, K., Zahedi, F.: The measurement of web-customer satisfaction: an expectation and disconfirmation approach. Inf. Syst. Res. **13**(3), 296–315 (2002)
18. Abbey, J.D., Kleber, R., Souza, G.C., Voigt, G.: The role of perceived quality risk in pricing remanufactured products. Prod. Oper. Manag. **26**, 100–115 (2017)
19. Oliver, R.L., Rust, R.T., Varki, S.: Customer delight: foundations, findings, and managerial insight. J. Retail. **73**(3), 311–336 (1997)
20. Kano, N.: Attractive quality and must-be quality. Hinshitsu (Qual. J. Jpn. Soc. Qual. Control) **14**, 39–48 (1984)
21. Anderson, E.W., Mittal, V.: Strengthening the satisfaction profit chain. J. Serv. Res. **3**(2), 107–120 (2000)
22. Kahneman, D., Tversky, A.: Prospect theory: an analysis of decisions under risk. Econometrica **47**(2), 263–291 (1979)
23. Homburg, C., Koschate, N., Hoyer, W.D.: The role of cognition and affect in the formation of customer satisfaction: a dynamic perspective. J. Mark. **70**(3), 21–31 (2005)
24. Rust, R.T., Oliver, R.L.: Should we delight the customer? J. Acad. Mark. Sci. **28**(1), 86–94 (2000)
25. Cadotte, E.R., Turgeon, N.: Dissatisfiers and satisfiers: suggestions from consumer complaints and compliments. J. Consum. Satisf. Dissatisfaction Complain. Behav. **1**(1), 74–79 (1988)
26. Bartikowski, B., Llosa, S.: Customer satisfaction measurement: comparing four methods of attribute categorisations. Serv. Ind. J. **24**(4), 67–82 (2004)
27. Schvaneveldt, S.J., Enkawa, T., Miyakawa, M.: Consumer evaluation perspectives of service quality: evaluation factors and two-way model of quality. Total Qual. Manag. **2**(2), 149–161 (1991)
28. Matzler, K., Sauerwein, E.: The factor structure of customer satisfaction: an empirical test of the importance grid and the penalty-reward-contrast analysis. Int. J. Serv. Ind. Manag. **13**(4), 314–332 (2002)
29. Busacca, B., Padula, G.: Understanding the relationship between attribute performance and overall satisfaction: Theory, measurement and implications. Mark. Intell. Plan. **23**(6), 543–561 (2005)
30. Albayrak, T., Caber, M.: Prioritisation of the hotel attributes according to their influence on satisfaction: a comparison of two techniques. Tour. Manage. **46**, 43–50 (2015)

31. Matzler, K., Renzl, B., Rothenberger, S.: Measuring the relative importance of service dimensions in the formation of price satisfaction and service satisfaction: a case study in the hotel industry. Scand. J. Hosp. Tour. **6**(3), 179–196 (2006)
32. Mikulic, J., Prebezac, D.: Prioritizing improvement of service attributes using impact range-performance analysis and impact-asymmetry analysis. Manag. Serv. Qual. Int. J. **18**(6), 559–576 (2008)

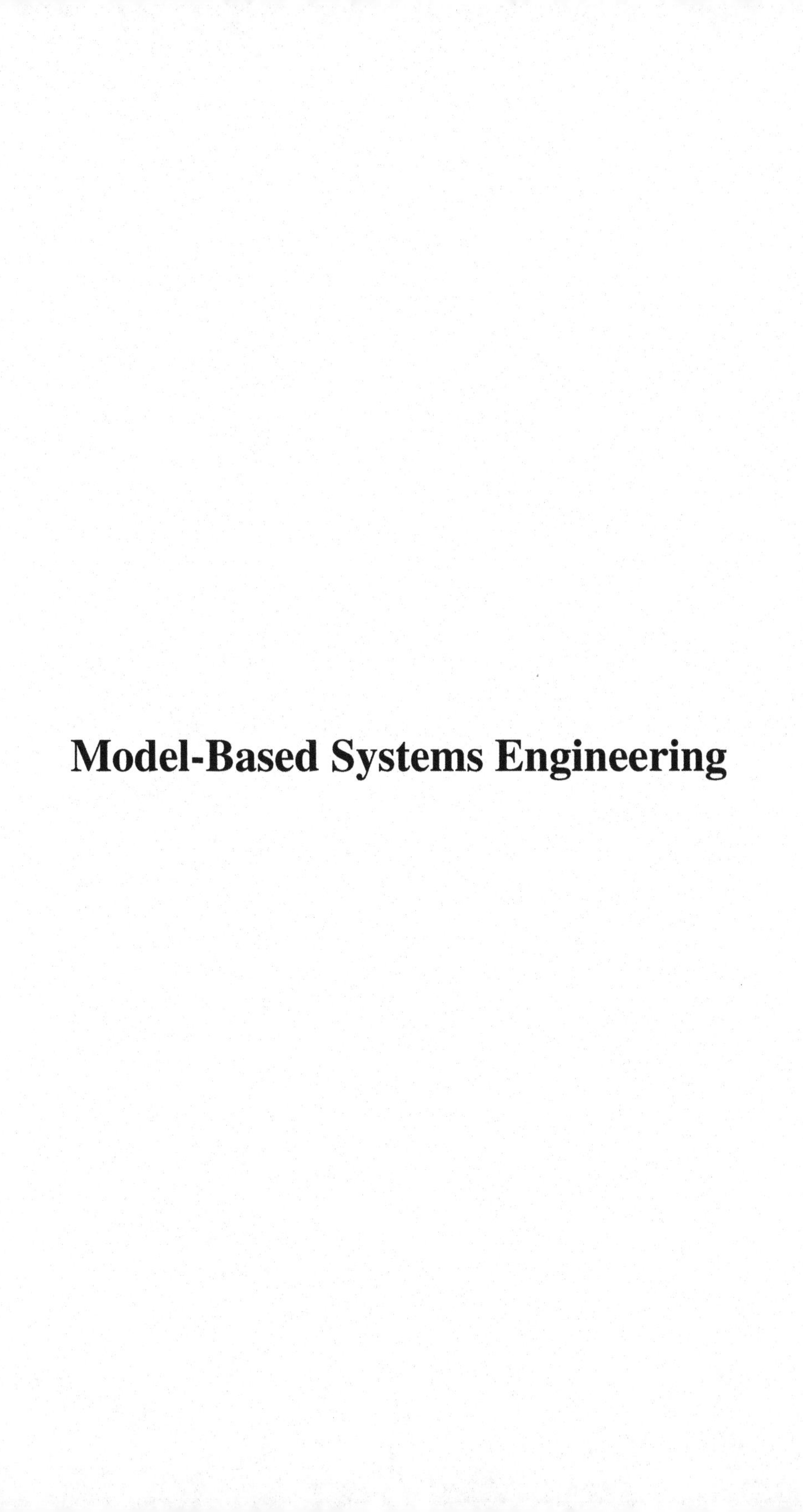

Model-Based Systems Engineering

Semantic Modeling Supporting Discrete Event Simulation for Aircraft Assembly Process

Xiaodu Hu[2], Jinzhi Lu[1](✉), Xiaochen Zheng[1], Rebeca Arista[3,4], Jyri Sorvari[5], Joachim Lentes[2], Fernando Ubis[5], and Dimitris Kiritsis[1]

[1] EPFL, 1015 Lausanne, Switzerland
jinzhi.lu@epfl.ch
[2] Fraunhofer IAO, 70569 Nobelstrasse, Germany
[3] Airbus SAS, 31700 Blagnac, France
[4] University of Seville, 41092 Seville, Spain
[5] Visual Components, 02600 Espoo, Finland

Abstract. Aircraft assembly is a complex process associated with different stakeholders, which strongly impacts the R&D cost for the entire aircraft lifecycle management. Traditional aircraft assembly process is designed based on a document-centric approach which cannot provide graphical notations for engineers to understand the entire assembly process. Moreover, it is difficult to make use of simulation to analyze the assembly process in an automatic way that increases the R&D cost for the assembly process design. In this paper, a semantic modeling approach is proposed to support aircraft assembly process formalism and performance analysis of the entire assembly process. First, semantic modeling language KARMA is used to develop meta-models based on the basic elements of discrete event simulation model for aircraft assembly process modeling. After KARMA models are developed, ontology model is generated from KARMA models for implementing the automatic simulation execution by using the discrete event simulation.

Keywords: Model-based systems engineering · Semantic modeling · KARMA language · Discrete event simulation · Aircraft assembly

1 Introduction

Aircraft assembly process design, operation and management is a complex systems engineering. Based on new type of equipment, such as modern milling and turning machine, it is possible to assemble the aircraft manufacturing parts with high accuracy from subsystems, with coordinate measuring machines to ensure the measurement of finished parts [24]. Though using such machines, the entire assembly process is still the most labor-intensive thus amount of stakeholders are required to support the assembly process by operating different machines and tools manually and automatically [10]. Different working tasks in the assembly process are conducted in a synchronous and asynchronous sequences, leading to that configuration of such working tasks with different operators is important because the way to implement such working tasks such as manual approach

J. Chen et al. (Eds.): KSS 2022, CCIS 1592, pp. 83–98, 2022.
https://doi.org/10.1007/978-981-19-3610-4_6

or automatic approach, can impact the R&D cost for the assembly. Thus, a model-based approach is required to support the assembly process design and simulation which enable to decrease the risks for budget and project control.

Model-based Systems Engineering (MBSE) was proposed since 1993 which is becoming a valuable novel trend in the field of systems engineering [18]. It is a formalized modeling approach to support systems requirement, design, analysis, verification and validation actives across the entire lifecycle. Currently, MBSE is widely used to support aircraft design and production. Architecture models enable to provide a set of standardized graphical notations to represent the aircraft system and assembly process. Simulation models provide a verification approach to analyze the performance of systems and assembly process.

In this paper, a semantic modeling approach is proposed to support MBSE for assembly process design and verification. First, KARMA language (Sect. 3.1) is used to develop meta-models for assembly process and construct models in order to represent the assembly operations using graphical notations. Then GOPPRRE (Graph-Object-Relationship-Role-Point-Property-Extension) ontology models (Sect. 3.2) are generated from KARMA models in order to support automatic generation of discrete event simulation (DES) models. Finally, the generated simulation models (Sect. 3.3) are executed by simulation engine, and simulation results for dynamic analysis of the assembly process.

The rest of the paper is organized as follows. The related works are discussed in Sect. 2. Then the proposed semantic modeling approach is introduced in details in Sect. 3. Section 4 elaborates how semantic model and discrete event model are integrated, as well as the details in discrete simulation. A case study is used to represent and discuss about how the approach supports aircraft assembly process modeling and simulation in Sect. 5. Finally, we offer the conclusions in Sect. 6.

2 Related Works

2.1 Assembly Process Modeling

Simple representation of an assembly process can be done in worksheets like Microsoft Excel/Project with Gantt-diagram, which can be used as spreadsheet for non-engineering people. A general approach to develop an assembly system in mathematical model is using petri-net (PN), for example, a timed petri nets to model an assembly process system [11], and a PN model with weighted arcs for the design and performance evaluation of flexible assembly system [26]. Besides, Semantic web technology modeling language, e.g. Ontology Web Language (OWL) can also be leveraged to represent an assembly system to achieve various levels of interoperability, which is based on Open Assembly Model and their previous Unified Modeling Language (UML) [12]. In addition, Business Process Model and Notation (BPMN) extensions are proposed for manufacturing domain including assembly process, however, the standard BPMN cannot model all details of assembly process [1]. Furthermore, the System Modeling Language (SysML) can be customized to create a domain-specific language to model a discrete event system [4]. Apart from those modeling approaches, other methods have also been explored, such as value Stream Mapping to model a complex production system including assembly process [7], an object-oriented aircraft assembly modeling approach [28] and graph-based assembly system modeling [8]. Particular in aerospace industry, an aero-structure aircraft assembly line model is proposed and mapped to CATIA V5 [20].

2.2 Discrete Event Simulation and Some Applications

In general, DES is dealing with a typical queueing model [3], concerning the future event list. Among others, it is usually used as a tool to evaluate the performance of a queueing system. In this paper, a non-exhaustive investigation is conducted to present a variety of DES applications and their application domains. As an operational research technique, DES has the advantage of modelling complex systems at the individual level, and accordingly numerous applications in the health care domain are implemented [29]. Besides, DES applications can also be found in the domains of computer and communication networks ([2, 6]). In the domain of supply chain and logistics, DES application is also widely used [25]. In the context of engineering and manufacturing, for example, product-service system development [27], automobile assembly layout pant [13], chemical plant development in process industry [23], maintenance scheduling in semi-conductor manufacturing [22], aircraft spare part management [17], etc.

2.3 Integration of Modeling and Simulation for Assembly Process

Early research has used SysML to integrate system modeling and simulation modeling, in order to resolve the issue of using different engineering tools [14]. An example of integration incorporating DES is to apply an electronic assembly system model as a customized SysML into Arena [5]. Besides, concerning the integration in the aerospace industry, a product, process and resource model in Delmia has been used, which interconnects related data including CAD data and enables the simulation of product performance, as well as perform network analysis for assembly operations concerning cycle-time and critical path [9]. It is an integrated approach from conceptual development to analysis conduction, which is, however, still with some gaps to a comprehensive multidisciplinary model integration. Moreover, using ontology to integrate modeling and simulation is also existing: a practical way of integrating ontology with simulation [21] is to adopt a formal system modelling language, e.g. SysML, to create an ontology with formal system description, and then conduct model transformation into conceptual models for simulation, where appropriate meta-models are fundamental. Nevertheless, there is no standardized meta-model for simulation.

Recently, another endeavor of semantic integration of models is to use a meta-meta model including Graph, Object, Point, Property, Role, and Relationship with Extensions (GOPPRRE) [19] to support model-based system engineering. It aims to tackle the aforementioned challenges due to its comprehensive representation capability for different modelling languages, as well as its generality as a meta-model. This paper adopts this GOPPRRE approaches for the semantic modeling.

3 Semantic Modeling Supporting Discrete Event Simulation Model

As shown in Fig. 1, a semantic modeling approach is proposed to support assembly process modeling and simulation. KARMA language (Sect. 3.1) is first used to develop meta-model and models for assembly process in an architecture modeling tool Meta-Graph 2.0 [18]. Then, through the GOPPRRE ontology model generator, GOPPRRE ontology models (Sect. 3.2) [19] are generated in order to support automated simulation model generation. Finally, an object-oriented discrete event simulation model is introduced (Sect. 3.3), which can be instantiated through the GOPPRRE ontology, and then be used to simulate the performance of the assembly process.

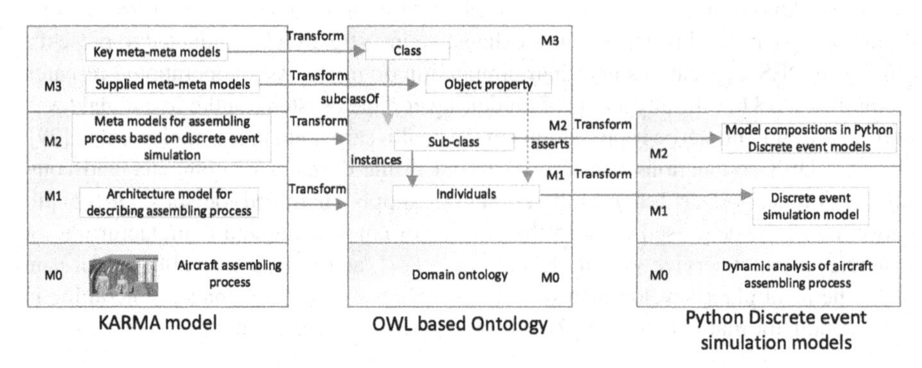

Fig. 1. Overview of the semantic modeling approach

3.1 KARMA Language Supporting Assembly Process Modeling

KARMA (Kombination of ARchitecture Model specificAtion) language is a semantic modeling language proposed by EPFL, KTH, Beijing Institute of Technology, etc. [18]. It enables to formalize multiple system architecture languages, such as SysML, BPMN, and etc. Moreover, KARMA models can be transformed into GOPPRRE ontology models (introduced in Sect. 3.2). As shown in Fig. 1, the KARMA syntax is designed based on a M0-M3 meta-object facility (MOF) framework. Meta-meta models include graph, object, relationship, role, point and property as shown in Table 1, which is considered as the most powerful meta-meta models for supporting the domain specific modeling language construction [15]. Meta-meta models are used to develop meta-models, which are considered as elements of assembly process, such as tasks, resource, etc. The meta-models are defined based on the meta-model of discrete event simulation model in order to realize the automatic generation of simulation models through the ontology models. Based on meta-models, assembly process models are developed. Based on the MOF framework, syntax of KARMA is designed and used to construct assembly process as shown in Fig. 2.

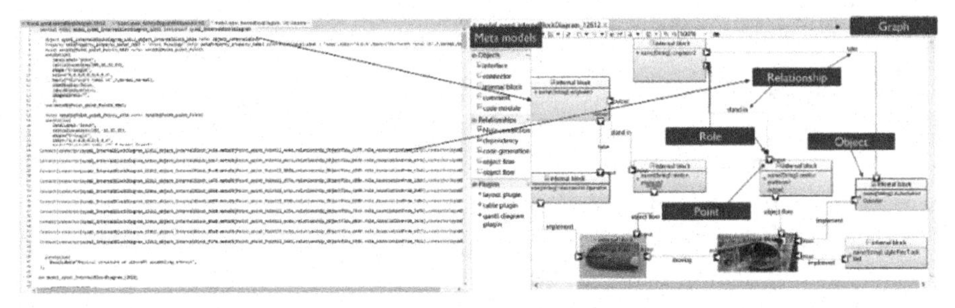

Fig. 2. KARMA models based on GOPPRR meta-models

Table 1. Key concepts when developing KARMA models

Meta-meta models	Description
Connector	It defines the connection rule between one role, relationship, and object in one end of each connection
Graph	It defines a meta-model which represent an assembly process of aircraft
Object	It defines a meta-model which represent an element of assembly process of aircraft, such as task
Relationship	It defines a meta-model which represent a connection of assembly process of aircraft
Role	It represents one end of a connection between model compositions
Point	It refers to a meta-model representing a port in each model composition
Property	An attribute for developing other meta-models
Language	A architecture modeling language constructed by other meta-models

3.2 GOPPRRE Ontology for Automatic Discrete Event Simulation

With the embedded GOPPRRE ontology model generator in MetaGraph 2.0, the KARMA models can be generated into GOPPRRE ontology model.

As shown in Fig. 1, the ontology entities and their topology were formalized based on the M0–M3 modeling framework whose class and object properties. The ontology entity enables the representation of a model structure as shown in Fig. 3. In order to represent the model structure, the graph individual is defined as one model including model compositions and the connections among them. Each graph individual (model structure) *has (object property)* object individuals which represent model compositions and relationship individuals which represent connections between object individuals. Each relationship individual *has (object property)* two role individuals which represent the two ends of the relationship individual. Each object individual *has (object property)* the point individuals which are defined as the port in the related objects. Two connector individuals are formalized to create a binding among role individual, relationship individual, and point individual. Based on the two connectors, the connection between object individual1 and object individual2 is created through relationship individual1.

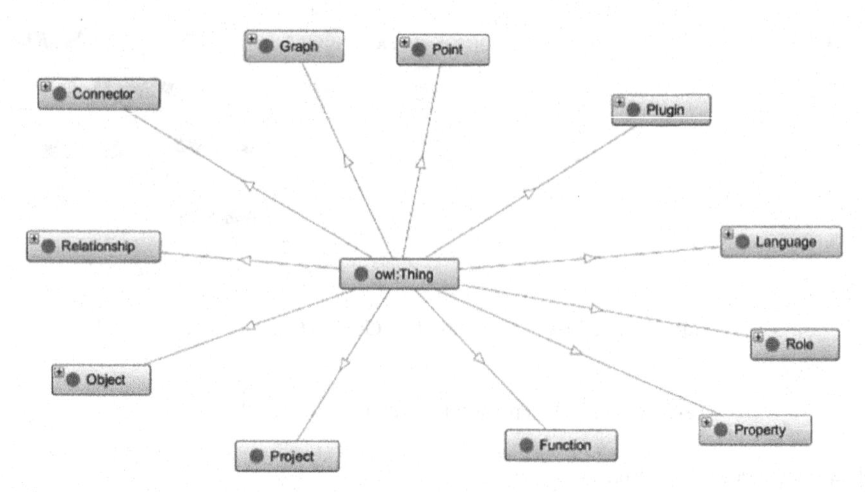

Fig. 3. KARMA models based on GOPPRRE meta-models

Each non-property individual *has (object property)* their own property individuals to define their attributes.

3.3 Discrete Event Simulation Model

Inheriting the aircraft assembly system model, i.e. the process and resource knowledge units [20], am object-oriented discrete event simulation model is developed in Python. A part of the developed classes of the model, their properties and methods are described in Fig. 4.

A *Scenario* creates a DES environment and presents a possible design option. Every *Scenario* instance created is unique to each other. The *Architecture* of each *Scenario* could be different due to the fact that some *Operations* can actually be parallelized.

An *Architecture* describes the relations among all involved *Operations*. Each *Scenario* owns only one *Architecture*, and it is a critical part of discrete even simulation model that drives the simulation engine.

Process and *Task* are at an equivalent level that represents an assembly process. These two classes are, however, to be differed apart. *Process* basically includes all process data from design side and is static. However, *Task* stands for a generated instance during simulation execution. Assume that the simulation is configured to simulate the assembly station for a year, in the time-based simulation environment, numerous fuselage assembly tasks could be done, and each *Task* tends to have different performance, e.g. different task lead-time and man hours. Note that the difference between *Process* and *Task* origins from the difference between data of design and data generated by simulation. A *Process*/*Task* consists of a series of *Operations*.

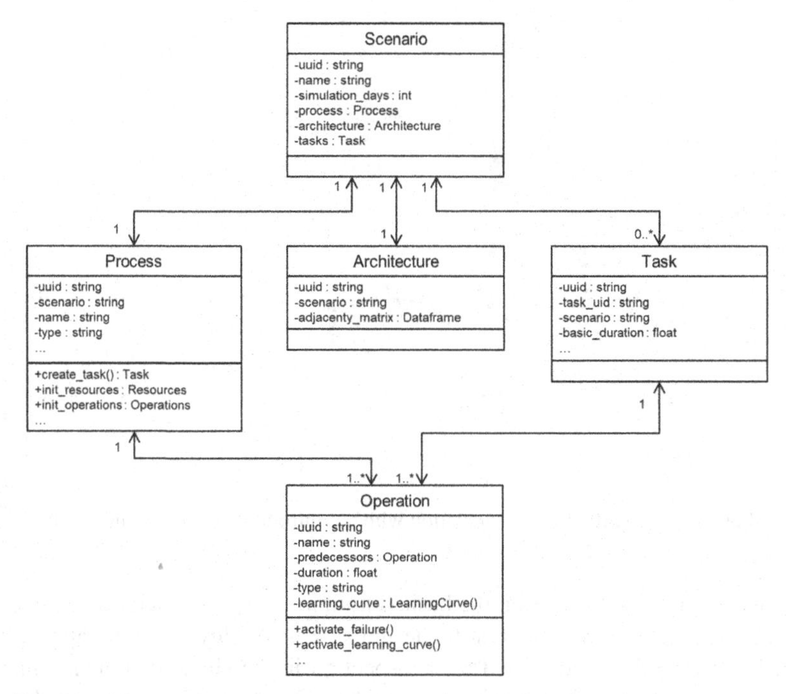

Fig. 4. A discrete event simulation model representing an assembly process (partly)

Operation is the lowest but fundamental unit in discrete event simulation model. As shown in Fig. 4. To avoid misunderstanding between different terms like assembly station, assembly process and the sub-processes under one process, the name *Operation* is given. The *Operation*, which is the lowest level of sub-process defined for simulation, should be distinguished from *Process*, which is a set of *Operation* that aims for assembling certain product.

3.4 Overall Equipment Effectiveness in Discrete Event Simulation Model

During simulation, operation instances and task instances are generated to virtually perform assembly works until the defined environment time limit. Each time an instance is created, the values of instance properties can vary. As depicted in Fig. 5, four different types of time are considered. They are useful operation time, net operation time, net running time and production time. They are considered as a simplified overall equipment effectiveness (OEE) model, which can be composed to quality rate, performance rate and availability rate.

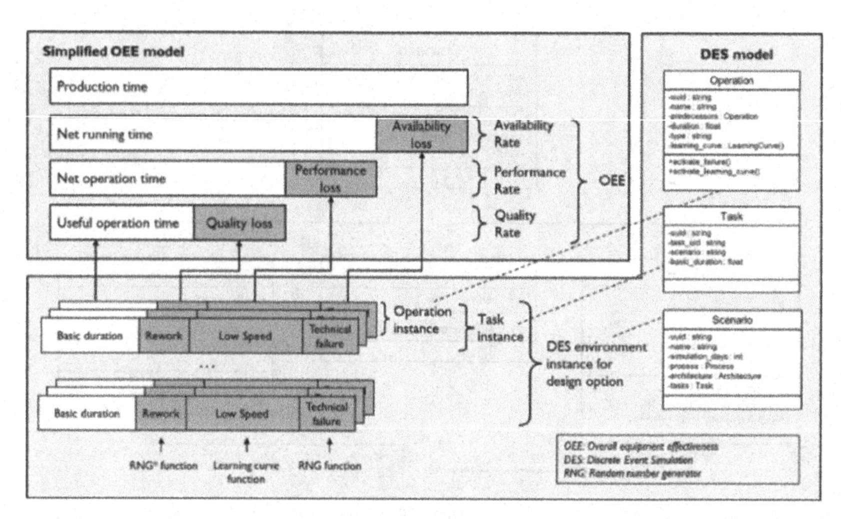

Fig. 5. A simplified OEE calculation with the introduction of learning curve

In this simplified OEE model, the basic duration defined in Operation class equals to the duration value retrieved from the design side. A *quality loss* can happen, when the needs for rework are identified after inspection in the simulation environment. It is triggered by a random number generator (RNG) function, where the probability of rework accords to the cumulative distribution function of normal distribution. A *performance loss* is caused by low speed. The low speed is determined by the efficiency of an operational work, which is simulated by a learning curve function that can be defined by user. An *availability loss* occurs when breakdown of machine or failure of jigs & tools happens, and the RNG function is the same like for rework. With these models, the OEE of the assembly process design can be simulated, in a defined range of simulation length. The incorporation of the simplified OEE model in DES model is depicted in Fig. 5.

4 Integrating Semantic Model and Discrete Event Simulation

In this section, the process of integrating discrete event simulation with the semantic model, i.e., GOPPRRE ontology, is elaborated. It mainly includes three steps: 1) From ontology (OWL) into an intermediate format, 2) From the intermediate into discrete the event simulation model and 3) Execution of discrete event simulation with loaded model (Fig. 6).

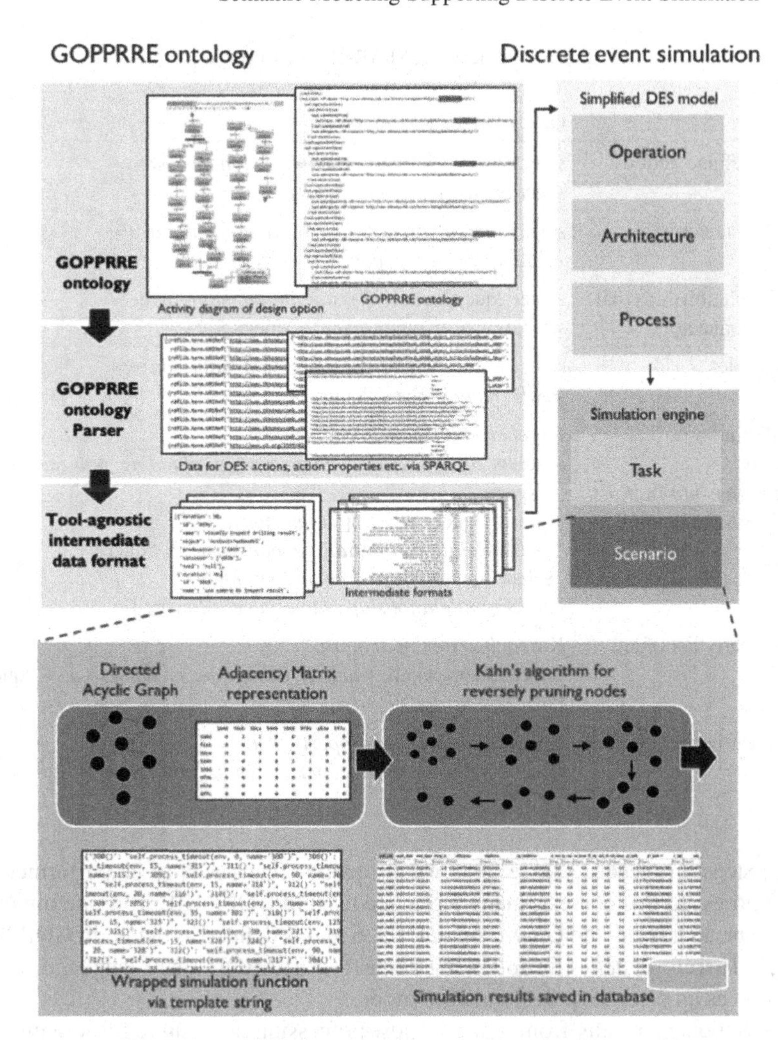

Fig. 6. From OWL via simulation model to queueing model

4.1 Data and Information Transformation from GOPPRRE Ontology

GOPPRRE ontology is intrinsically organized as a Resource Description Framework (RDF) format. To fetch data and information stored in RDF graphs, SPARQL[16] as a query language is used. In this paper, a python-library *rdflib* is used with its plugin of SPARQL to query the needed data and information for instantiating the discrete simulation models.

In the initial step, a name space needs to be configured, as well as the name space especially for discrete simulation, e.g. *desSimulation*. With the pre-set name spaces, certain SPARQL query can be written. Besides, the GOPPRRE ontology stores the needed data in a manner that is represented *graphs*, *objects*, *points*, *properties*, *roles*, and *relationships*. Thus, those data and information cannot be directly fetch. Instead,

Table 2. Query via SPARQL and purpose

Purpose	Query language example
Fetch all defined actions	select ?query where { *<simulation_namespace>* *<action_namespace>* ?query }
Fetch all connectors	select ?query where { *<simulation_namespace>* *<connector_namespace>* ?query }
Fetch all relationships that connects connectors	select ?query where { *<simulation_namespace>* *<relationship_namespace>* ?query }
Fetch all roles	select ?query where { *<simulation_namespace>* *<role_namespace>* ?query }
Fetch all bindings: – *bindings* – *linkRelationshipAndRole* – *linkFromRelationship*	*For each connector in connectors*: – select ?query where { *connector_id <binding_namespace>* ?query } – select ?query where { *connector_id* *<linkRelationshipAndRole_namespace>* ?query } - select ?query where { *connector_id* *<linkFromRelationship_namespace>* ?query }
Fetch property list of an action	*For each action in actions*: select ?query where { *action_id <property_namespac >* ?query }
Fetch all properties of an action	*For each property in property_list*: select ?query where { *property <value_namespace>* ?query }

multiple steps are needed. Table 2 shows all queries that needs to be performed and their purposes. In this context, many terms are used in GOPPRRE. But one important term is to be addressed: the *Operation* in DES model is defined as *action* in GOPPRRE ontology. Therefore, *action* in following context of SPARQL query languages can be understood as an *Operation* in simulation mode.

With the query results from Table 2, post-processing are required to organize the data and information. Some mapping dictionaries are established in post-processing: 1) mapping between role and action and 2) mapping between role and relationship. Then, the relation between different roles, to be more specific, *ObjectflowFrom* or *ObjectflowTo*, can be identified by using the relationships query result and the mapping between role and relationship. Later, the relation between actions can be synthesized with both role-action mapping and role-role relation. Until this step, all actions and their relations are resolved.

As shown in the last two lines in Table 2, based on the retrieved actions, all properties and their values of an *action* can be retrieved, for instance, the name, predecessors successors, duration, etc., and then saved in an intermediate format.

4.2 Discrete Event Simulation Execution

In order to execute DES, the DES model needs to be initialized in the first place. Then, the models are fed into DES environment and the simulation is executed.

Initializing DES Model: The first step is to instantiate the class *Operation* for the retrieved data and information queried from GOPPRRE ontology, which represents a design option including process and architecture information. As long as all operation instances are created, the *Process* class takes the list of operation instances as input and instantiates a process that contains all operation related properties and values for this design option. At the meantime, the class *Architecture* iterates the list of operation instances, and creates an adjacency matrix that represents the assembly operation topology. Until now, the initialization of DES model for one design option is finished.

As shown in Fig. 6, the simulation engine takes the process and architecture instance as input for simulation. Note that the process instance already includes all operation instances. Because the adjacency matrix can represent a directed acyclic graph with vertex (current operation) and relation (from predecessor operation to current operation), the core idea of Kahn's algorithm is used to prune the graph saved in the adjacency matrix, resulting in identified relation between each pair of operations. These operations are then wrapped into an executable string format for simulation engine by using a string template: from the last operation backwards to the first operation, if two parallel successor operations are identified, they are wrapped into a *parallel_process* function by specifying *AllOf*. It means, only after all of the parallelized operations are finished, the assembly process state can then move forward to the next operation. For the case when two sequential operations (one after another) are identified, they are then wrapped into a *linear_process* function without any specification, because the next operation will starts automatically after the previous operation is finished.

DES Environment: A python-library SimPy [30] as DES engine is used. After the string codes with executable functions for each operation are prepared, all functions are then executed by passing over the duration of each operation into the corresponding function. We configure the simulation length to be 365 days. However, each iteration of going through all operations might not need that long. Thus, numerous *Task* instances are generated according to the process and architecture instance. Each task instance represents the finalization of the assembly task – the fuselage joint process. In this context, the discrete event simulation environment has an environment clock starting from 0 min. The clock is advanced to the next environment time, when an operation is finished, and it the clock is ended until the 262.800 min. This number is the simulation length in DES environment, and it indicates the sum of minutes in 365 working days per year, with 12 working hours per day. In DES, each operation stands for two events: 1) operation starts and 2) operation ends. Nevertheless, in this paper, the events are not exclusively defined, because it is that important to record the events of operation start and end, as the main purpose of this paper is to validate the integration approach rather than performing further optimization which needs the event list. Furthermore, the learning curve and random number generator (RNG) to trigger random failure that triggers time delays are also included during simulation execution (Fig. 5).

4.3 Discrete Event Simulation Results

The discrete event simulation results are directly stored in a database during the simulation execution. The results record currently the task-level information, for example, the task id, start date, end date, day of task start, the lead-time, availability and performances rate and so on. A further traceability of operation information is also possible. The simulation data is further used for visualization.

5 A Case Study

5.1 Introduction Aircraft Assembly System Design

The use case of aircraft assembly system design is based on a pilot project in an European funded project QU4LITY. The case refers to an assembly station that performs fuselage assembly (a in Fig. 7). In this station, the rear and front fuselage of an aircraft is joint together, starting from setting up the assembly environment including jig-in, setting up crane for rail for light Flextrack robotics, via drilling activities, riveting buttstraps, etc. and ending up with the inspection operation and jig-out operation. In the early phase of an aircraft program, many engineering activities are included for the industrial system design, for instance, system architecture modelling, system performance simulation and evaluation, design option trade-off and system design decision making. Different modelling languages can be used for architecture modelling at different levels based on MOLFP (b in Fig. 7). Gantt diagram and PERT are applied to describe assembly process sequence which also contains some information like operation name, operation duration, etc. (c in Fig. 7). The top-level industrial requirements and their variants, which are significant to measure and evaluate the developed system designs, are created, maintained and manually tracked via a SysML tool (d in Fig. 7).

a) Orbital joint and fuselage assembly process

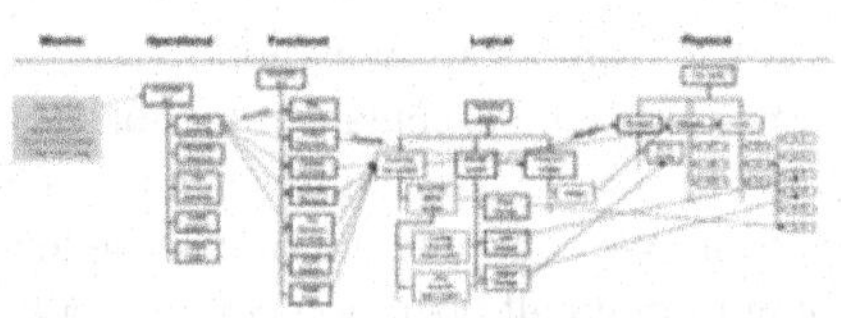

b) Architecture modelling based on MOLFP

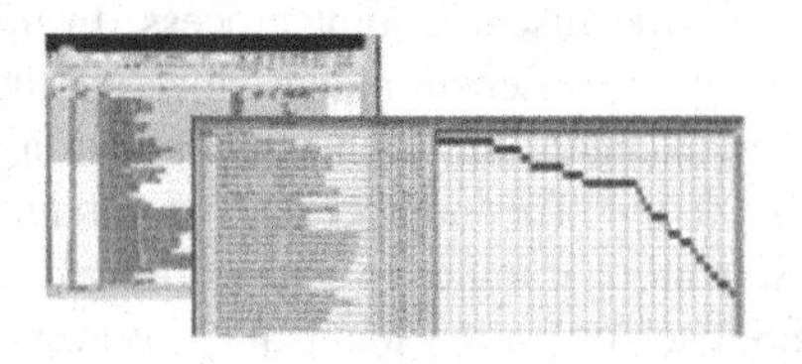

c) Process description using Gantt digram

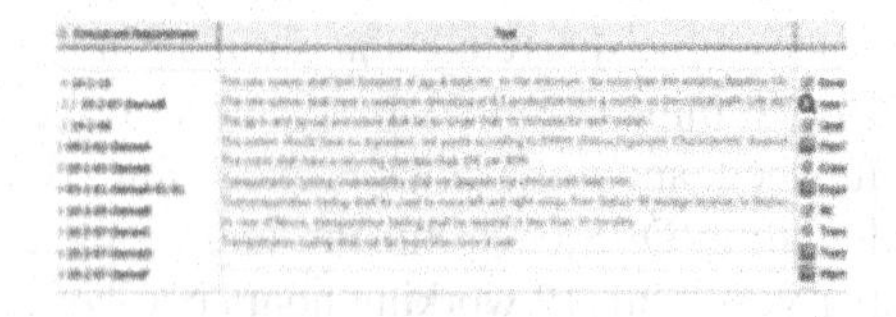

d) Requirements definition in SysML

Fig. 7. Steps of case study for semantic modeling supporting discrete event simulation

To integrate the dispersed engineering activities in a MBSE formalism, the proposed approach of semantic modeling that can represents interdisciplinary and multi-faceted models and methods, is applied. The implementation is introduced in Sect. 5.2.

5.2 Case Implementation

The purpose of this implementation is to validate that the semantic modelling approach, which is capable of representing different modelling languages, can serve as the semantic core for other services in later engineering activities, i.e. simulation and evaluation. In this paper, only one simulation method is used for approach validation—the discrete event simulation approach concerning OEE mentioned in Sect. 3.3 and 3.4, as it fits the need in this project at most. Accordingly, the integration of further modelling language and other evaluation methods via the proposed approach are not the focus in this paper.

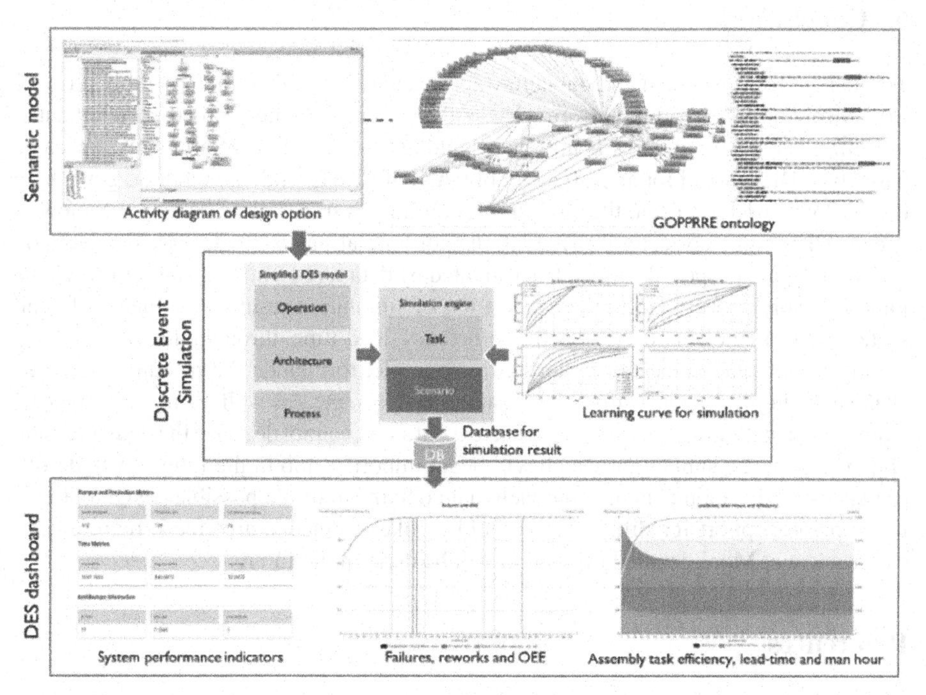

Fig. 8. Steps of case study for semantic modeling supporting discrete event simulation

As illustrated in Fig. 8, the discrete event diagram of the assembly process is modeled in MetaGraph. Via KARMA language, it can be represented in GOPPRRE ontology in the first step. Next, the ontology in OWL format is taken by the DES tool. After data and information transformation from GOPPRRE ontology into DES model, the simulation engine receives the DES model including a process instance and an architecture instance, and conducts the parsing of DES and creates the simulation string template for simulation engine SimPy. The environment clock is advanced based on the topology of the assembly process and their operation duration, which are determined not only by basic duration from design option, but also by random delays caused by failure and the operational efficiency concerning a learning curve. While the simulation is executing, the simulation results is saved in a database SQLite. Finally, the simulation results are visualized via a web-based dashboard, which presents the performance indicators of the designed system and illustrates the simulation records, such as the history of failure and rework times within the simulation length, impacts on OEE, as well as the lead-time, man hour changing of each task.

Figure 8 depicts the end-to-end implementation from modeling of assembly process, via discrete event simulation for performance analysis, upon the visualization for support decision. The implementation indicates the potential of meta-model that can be leveraged for further integration of cross-disciplinary engineering methods and models, thus achieving the interoperability in aircraft assembly design.

6 Conclusion

This paper proposes a semantic modeling approach to support aircraft assembly process formalism and its dynamic performance analysis of an assembly process. A semantic modeling language KARMA is introduced and used to develop meta-models, which is based on a DES model for aircraft assembly. A GOPPRRE ontology model is generated by KARMA models. Then, the GOPPRRE ontology is transformed to the DES model, and the DES engine executes the simulation with simulation results stored in database. In a case study, the proposed approach is validated by data representing an aircraft fuselage joint assembly process, and the visualization of simulation results illustrates different system performance indicators, such as efficiency, lead-time, throughputs, etc.

Further potentials can be explored, for example, to include the resources for cost analysis including cost sensitivity and Pareto analysis, etc., as well as the performance requirement validation. Moreover, the current ontology cannot describe the dynamic data obtained from the simulation which will be an important job in the future. A trade-off can also be built upon existing approach that contains multiple possible design options for decision support for different stakeholders like engineers, architects, requirement engineers, etc. More detailed evaluation will be done in the future.

References

1. Abouzid, I., Saidi, R.: Proposal of BPMN extensions for modelling manufacturing processes. In: 2019 5th International Conference on Optimization and Applications (ICOA), pp. 1–6. IEEE (2019)
2. Alam, F.M., Mohan, S., Fowler, J.W., Gopalakrishnan, M.: A discrete event simulation tool for performance management of web-based application systems. J. Simul. **6**, 21–32 (2012). https://doi.org/10.1057/jos.2011.8
3. Banks, J., Nelson, B.L., Carson, J.S., Nicol, D.M.: Discrete-Event System Simulation, 5th edn., p. 624. Pearson, Upper Saddle River (2010)
4. Batarseh, O., McGinnis, L.F.: System modeling in SYsML and system analysis in Arena. In: Proceedings of the 2012 Winter Simulation Conference (WSC). IEEE, pp. 1–12 (2012)
5. Batarseh, O., McGinnis, L.F.: SysML to discrete-event simulation to analyze electronic assembly systems. Simul. Ser. **44**, 357–364 (2012)
6. Bonaventura, M., Foguelman, D., Castro, R.: Discrete event modeling and simulation-driven engineering for the ATLAS data acquisition network. Comput. Sci. Eng. **18**, 70–83 (2016). https://doi.org/10.1109/MCSE.2016.58

7. Braglia, M., Carmignani, G., Zammori, F.: A new value stream mapping approach for complex production systems. Int. J. Prod. Res. **44**, 3929–3952 (2006). https://doi.org/10.1080/00207540600690545
8. Breckle, T., Kiefer, J., Rudolph, S., Manns, M.: Engineering of assembly systems using graphbased design languages. In: Proceedings of the International Conference on Engineering Design, ICED 1 (2017). Engineering of assembly systems using graph-based design languages
9. Butterfield, J., et al.: An integrated approach to the conceptual development of aircraft structures focusing on manufacturing simulation and cost. In: Collection of Technical Papers - AIAA 5th ATIO and the AIAA 16th Lighter-than-Air Systems Technology Conference and Balloon Systems Conference, vol. 1, pp. 506–515 (2005). https://doi.org/10.2514/6.2005-7355
10. Chen, Z., Tang, J.: Aircraft digital assembly process design technology based on 3D Model. MATEC Web. Conf. **202**, 02004 (2018). https://doi.org/10.1051/matecconf/201820202004
11. Dejan, G.G., Musi, C.: Petri-net modelling of an assembly process system. In: Proceedings of the 7th International Ph.D. Workshop: Young Generation Viewpoint, p. 16 (2006)
12. Fiorentini, X., Gambino, I., Liang, V.-C., Rachuri, S., Mani, M., Bock, C.: An Ontology for Assembly Representation, Gaithersburg, MD (2007)
13. Halim, N.N.A., Shariff, S.S.R., Zahari, S.M.: Modelling an automobile assembly layout plant using probabilistic functions and discrete event simulation. In: Proceedings of the International Conference on Industrial Engineering and Operations Management (2020)
14. Huang, E., Ramamurthy, R., McGinnis, L.F.: System and simulation modeling using SYSML. In: 2007 Winter Simulation Conference, pp. 796–803. IEEE (2007)
15. Kelly, S., Lyytinen, K., Rossi, M.: MetaEdit+ a fully configurable multi-user and multi-tool CASE and CAME environment. In: Constantopoulos, P., Mylopoulos, J., Vassiliou, Y. (eds.) Seminal Contributions to Information Systems Engineering, pp. 109–129. Springer, Heidelberg (2013). https://doi.org/10.1007/978-3-642-36926-1
16. Kumar, N., Kumar, S.: Querying RDF and OWL data source using SPARQL. In: 2013 4th International Conference on Computing, Communications and Networking Technologies, ICCCNT 2013, pp. 1–6. IEEE (2013)
17. Li, H., Zhu, Y., Chen, Y., Pedrielli, G., Pujowidianto, N.A.: The Object-Oriented discrete event simulation modeling: a case study on aircraft spare part management. In: 2015 Winter Simulation Conference (WSC), pp. 3514–3525. IEEE (2015)
18. Li, Z., Lu, J., Wang, G., Feng, L., Broo, D.G., Kiritsis, D.: A bibliometric analysis on model-based systems engineering. In: 2021 IEEE International Symposium on Systems Engineering (ISSE), pp. 1–8. IEEE (2021)
19. Lu, J., Ma, J., Zheng, X., Wang, G., Li, H., Kiritsis, D.: Design ontology supporting model-based systems engineering formalisms. IEEE Syst. J. **14**, 1–12 (2021). https://doi.org/10.1109/JSYST.2021.3106195
20. Mas, F., Ríos, J., Menéndez, J.L., Gómez, A.: A process-oriented approach to modeling the conceptual design of aircraft assembly lines. Int. J. Adv. Manuf. Technol. **67**, 771–784 (2013). https://doi.org/10.1007/s00170-012-4521-5
21. McGinnis, L., Huang, E., Kwon, K.S., Ustun, V.: Ontologies and simulation: a practical approach. J. Simul. **5**, 190–201 (2011). https://doi.org/10.1057/jos.2011.3
22. Scholl, W., Mosinski, M., Gan, B.P., Lendermann, P., Preuss, P., Noack, D.: A multi-stage discrete event simulation approach for scheduling of maintenance activities in a semiconductor manufacturing line. In: Proceedings of the 2012 Winter Simulation Conference (WSC), pp. 1–10. IEEE (2012)
23. Sharda, B., Bury, S.J.: Best practices for effective application of discrete event simulation in the process industries. In: Proceedings of the 2011 Winter Simulation Conference (WSC), pp. 2315–2324. IEEE (2011)

24. Soliman, M.S.E.A.: Assembly Process in Aircraft Construction (2017)
25. Tako, A.A., Robinson, S.: The application of discrete event simulation and system dynamics in the logistics and supply chain context. Decis. Support Syst. **52**, 802–815 (2012). https://doi.org/10.1016/j.dss.2011.11.015
26. Ullah, H., Bohez, E.L.J.: A Petri net model for the design and performance evaluation of a flexible assembly system. Assem. Autom. **28**, 325–339 (2008). https://doi.org/10.1108/01445150810904486
27. Weidmann, D., Maisenbacher, S., Kasperek, D., Maurer, M.: Product-Service System development with Discrete Event Simulation modeling dynamic behavior in Product-Service Systems. In: 2015 Annual IEEE Systems Conference (SysCon) Proceedings, pp. 133–138. IEEE (2015)
28. Xu, K., Wang, C., Zhang, W.: Object-oriented aircraft assembly model. In: Proceedings of the International Conference on Computer, Networks and Communication Engineering (ICCNCE 2013). Atlantis Press, Paris (2013)
29. Zhang, X.: Application of discrete event simulation in health care: a systematic review. BMC Health Serv. Res. **18**, 687 (2018). https://doi.org/10.1186/s12913-018-3456-4
30. Zinoviev, D.: Discrete Event Simulation. It's Easy with SimPy, pp. 1–18. PragPub (2018)

A Unified Model-Based Systems Engineering Framework Supporting System Design Platform Based on Data Exchange Mechanisms

Yuqiang Guo[1](✉), Jingbiao Wei[2], Guoding Chen[1], and Shiyan She[1]

[1] China Helicopter Research and Development Institute, Tianjin 300308, China
Linkation@126.com
[2] Army Aviation Research Institute, Beijing 101121, China

Abstract. The system model built with SysML has strong connections with requirements, multidisciplinary analysis, hardware and software development, verification, etc. A unified MBSE design platform is needed to support the dense and heterogeneous modelling information exchange. In this paper, four data exchange mechanisms to create a unified MBSE Design platform are studied and introduced, including modelling standard-based exchange, API-based exchange, multidisciplinary optimization tool-based exchange, and data bus-based exchange. A unified MBSE design platform framework is given in the end for reference. Compared with other frameworks, this solution uses the system modelling tool as a hub in the centre. It is flexible and fully utilises the existing modelling tools and data exchange specifications to save cost and time to build an MBSE design platform. A helicopter fuel system model as a case study demonstrates the feasibility of a unified MBSE modelling platform.

Keywords: MBSE · Modeling platform · Data exchange

1 Introduction

To take full advantage of MBSE and carry out model-based design, analysis, and verification throughout the system lifecycle, it is critical to establish a unified MBSE design platform, which includes a broad spectrum of modelling software that supports the system lifecycle development processes [1].

There are two ways to develop the platform, a new platform development or an integrated platform using existing software and hardware. The time and economic cost of new platform development are too heavy for companies, and it is challenging to transfer previous design experience to the new platform. So based on the existing design, software and hardware integration is a feasible and economical solution.

The SPIRIT Tool-chain framework supporting tool integration is based on OSLC specification and API development [2]. Nevertheless, this requires a deeper understanding of the software and software development, which will also need a lot of time and cost. Furthermore, on the other hand, some software has no APIs to support the integration.

J. Chen et al. (Eds.): KSS 2022, CCIS 1592, pp. 99–112, 2022.
https://doi.org/10.1007/978-981-19-3610-4_7

In response to these challenges, a unified MBSE design platform framework is proposed based on various data exchange mechanisms. The second part of the article analyzes four paths to create a unified MBSE Design platform. The third part of the article gives a possible unified MBSE design platform that appears to implement MBSE using the mainstream system and engineering tools. To demonstrate the feasibility of a unified MBSE modelling platform, we use the case of a helicopter fuel system model in the fourth part. Finally, we concluded at the end of this article.

2 Four Data Exchange Mechanisms to Create a Unified MBSE Design Platform

The system model built by SysML has strong connections with requirements, multidisciplinary analysis, downstream hardware and software development, verification, etc. The data connections between general engineering tools and the system modelling tool have been studied by Stanford and can be shown in Fig. 1 [3].

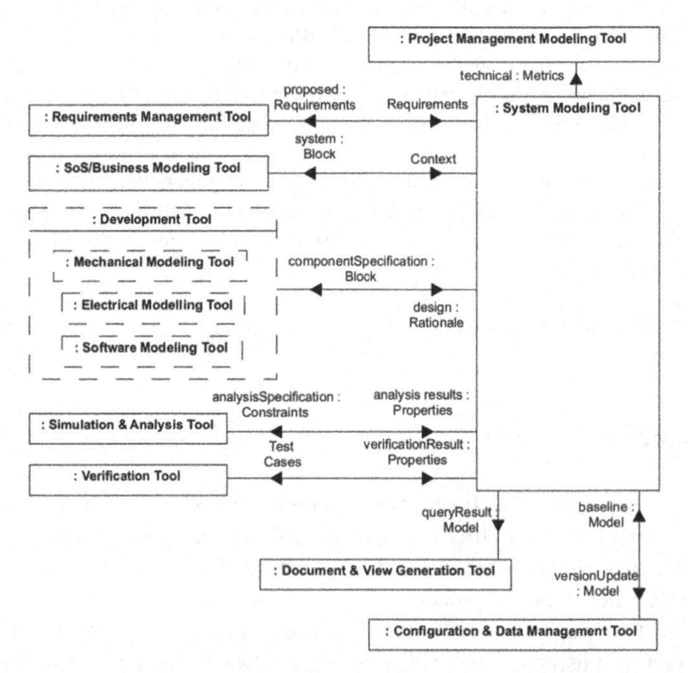

Fig. 1. Data connections between general engineering tools and the system modelling tool

The figure shows that the system modelling tool is like a hub of the engineering toolchain. Requirements information should be exchanged between the requirements management tool and the system modelling tool; Operation context information shall be exchanged between the SoS modelling tool and the system modelling tool; Component specification should be exchanged between the development tool and system modelling tool. Analysis results shall be imported into the system modelling tool; Test cases from the system modelling tool shall be imported into the verification tool. It is meaningful to study the message exchange mechanism between system modelling tools and other engineering tools.

In the MBSE design platform, when a model element in the system model is changed, possible data exchange may be needed to maintain the overall model consistency. The platform can be built in different ways using various data exchange mechanisms.

2.1 Modelling Exchange Standard-Based Exchange

(1) XMI, XML, and MOF combined metadata interchange

OMG released Meta-Object Facility (MOF) as a modelling and metadata repository standard [4]. Furthermore, 3WC released XML as a text-based language that uses tags to describe structured data [5]. XMI provides a set of rules to convert MOF meta-model UML profiles into a set of tags in XML.

OMG language families, including UML, SysML, and UAF/UPDM, can all be treated as UML profiles. This creates the conditions for data conversion from SoS/Business model to the system model and forms the foundation for a unified SoS-MBSE methodology.

The software modelling language is UML; the SysML model can be transformed to UML based on the XMI specification to perform detailed software functions and architecture design. The UML model can even be used to generate software code automatically (Fig. 2).

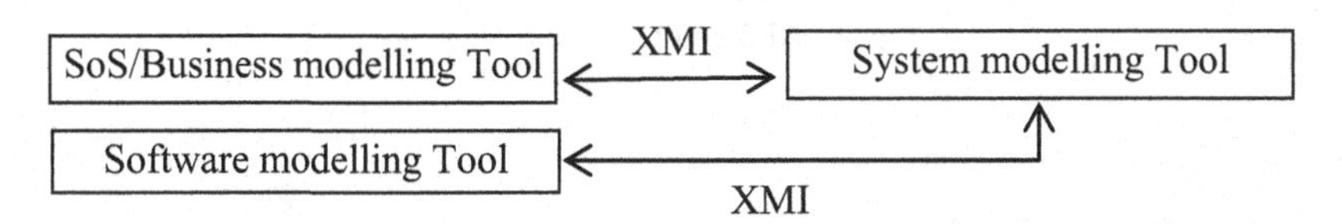

Fig. 2. XMI, XML, and MOF combined metadata interchange

(2) Sysphs for data exchange

SysML extension for physical interaction and signal flow simulation (SysPhs) extends SysML with additional information needed to model physical interaction and signal flow simulation independently of simulation platforms [6]. SysPhs toolbox includes SysML modelling elements that support OpenModelica and MATLAB Simulink. This builds a bridge between simulation tools and system modelling tools (Fig. 3).

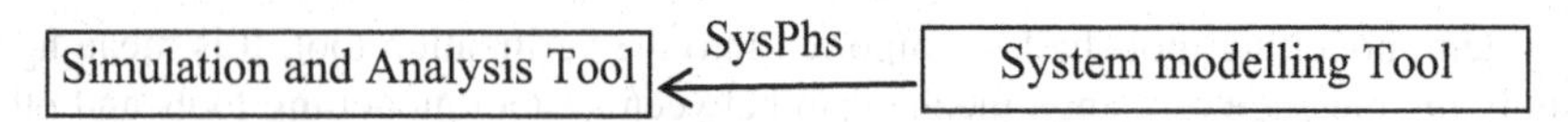

Fig. 3. Sysphs for data exchange

(3) Application Protocol 233 for data exchange

STEP, the Exchange of Product Model Data standard, is an international standard known as ISO 10303. It provides a mechanism to describe product data throughout a product's lifecycle, independent of any particular system [7]. AP233, Application Protocol 233 is a STEP-based data exchange standard that supports data exchange between different SysML modelling tools and CAD tools. Many other related work areas and related standards, together with AP-233, have been studied [8].

The structure and interface information in the system model provides a model-based specification that guides the detailed mechanical and electrical modelling (Fig. 4).

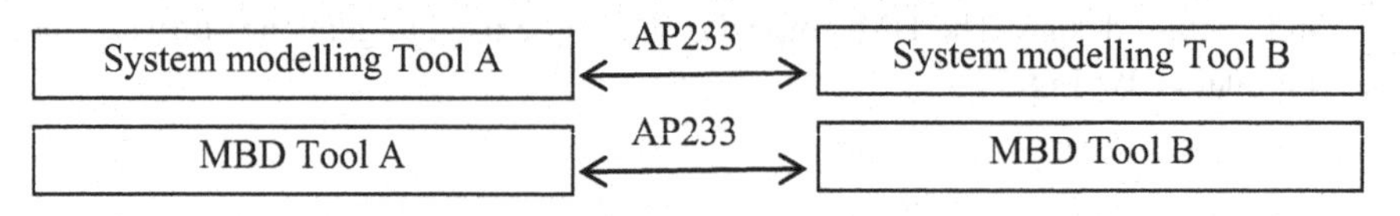

Fig. 4. AP233 for data exchange

(4) OSLC for data exchange

OSLC, Open Services for Lifecycle Collaboration, is a specification that creates a family of web services to make data exchange between different commercial softwares [9]. It provides a standard API that OSLC-compliant tools can request system model data, or conversely, system modelling tools request data from OSLC-compliant tools. Some modelling tool vendors have realized requirements and change management based on OSLC specifications until now.

This supports requirements synchronization between requirements management tools and system modelling tools (Fig. 5).

Fig. 5. OSLC for data exchange

(5) FMI for data exchange

FMI, Functional Mock-up Interface, is a specification that supports both model exchange and co-simulation [10]. It can be used in the system analysis and convert the simulation/analysis model into a Functional Mockup Unit(FMU) and simulate in the system modelling tools. Moreover, exporting SysML blocks as FMUs is also realized in engineering practices.

FMI is a widely accepted international standard that provides an effective way to bridge simulation tools and system modelling tools. Nevertheless, the FMU is a black box for the engineer, and the simulation model can not be changed while processing (Fig. 6).

Fig. 6. FMI for data exchange

(6) SysML-Modelica transportation for data exchange

OMG builds SysML-Modelica Transformation Specification to support the SysML model and Modelica model data exchange[11]. It provides another standard to link the simulation/analysis tool and the system modelling tool to leverage the two language strengths.

Modelica is a multi-physics domain simulation language and has developed rapidly in recent years. This provides a uniform modelling language for different simulation domains, and the simulation model is a white box that engineers can change the model while processing. Otherwise, it needs engineers to re-model, and the numerous existing model can not be used directly in the modelling environment (Fig. 7).

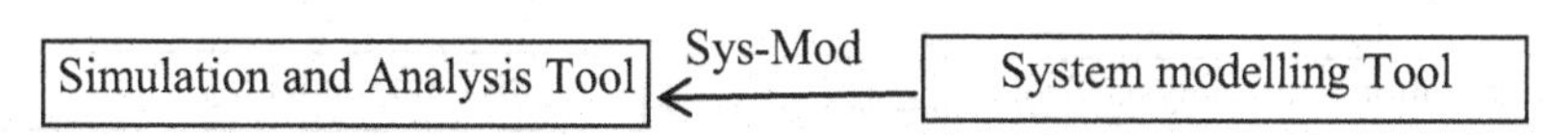

Fig. 7. SysML-Modelica transportation for data exchange

(7) ReqIF for data exchange

Requirements Interchange Format (ReqIF) is an XML file format built by OMG [12]. It provides the ability to import and export requirements between Requirements Management tools and SysML modelling tools from mainstream tool vendors.

Mainstream MBSE tools, like MagicDraw, Capella, EA, and Rhapsody support this file standard well (Fig. 8).

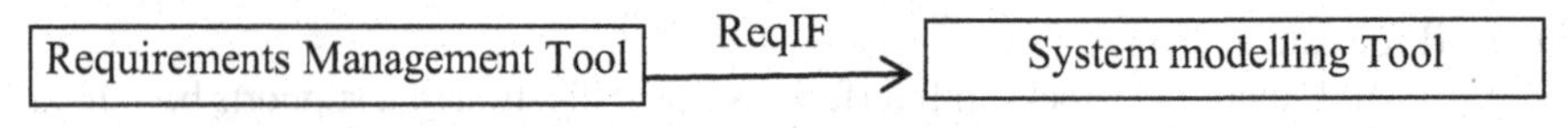

Fig. 8. ReqIF for data exchange

2.2 API-Based Exchange

Application programming interface (API) is another way to perform the data exchange between different tools. It is a quick, repeatable, reliable, but expensive way to achieve data exchange, it needs the software coding ability, and the engineers need to know precisely how the software works.

Suppose the engineers' team know the software well and there is no modelling exchange standard to support the data exchange. In that case, it is a qualified and alternative way to perform the data exchange between system modelling tools and other engineering tools. It is an efficient way to link Self-develop software to the system modelling tool (Fig. 9).

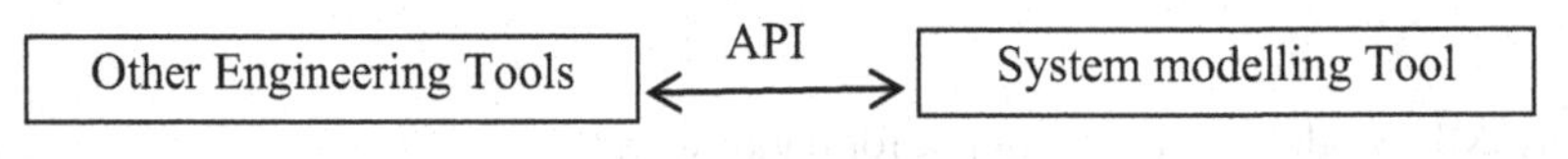

Fig. 9. API-based data exchange

2.3 Multidisciplinary Optimization Tool-Based Exchange

Multidisciplinary optimization tool like ModelCenter or Isight is a proper method to bridge the gap between System Model and Engineering Analysis Model. It provides users to graphically create simulation workflows to automate the execution of almost any modelling and simulation tools with system modelling tools.

On the other hand, a large number of optimization algorithms can be applied to the engineering process using this data exchange way (Fig. 10).

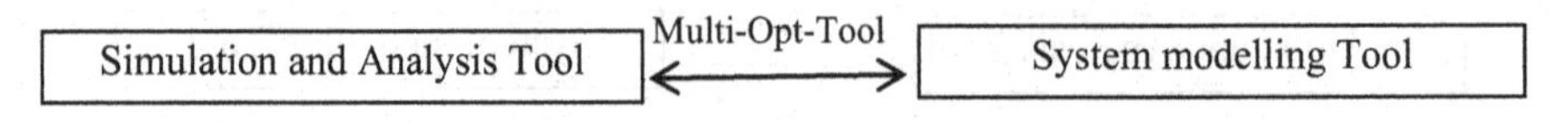

Fig. 10. Multidisciplinary optimization tool-based exchange

2.4 Databus-Based Exchange

Databus is a data-centric software framework for distributing and managing real-time data in an intelligent distributed system. It allows applications and devices to work together as one integrated system. This technology has been successfully used in the MBSE solutions [13] in the connection of System models and simulation models, especially for long-distance distributed simulation systems. It is a software-independent method to perform data exchange between most commercial software (Fig. 11).

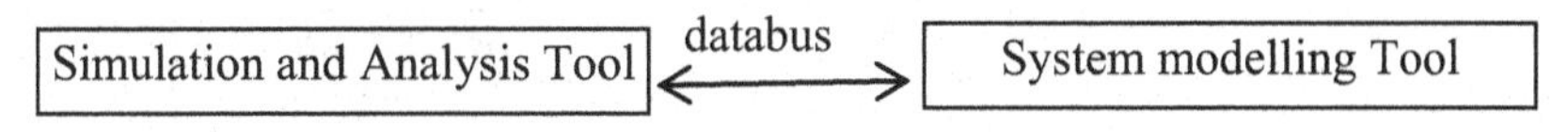

Fig. 11. Data bus-based exchange

3 A Unified Framework Supporting MBSE Design Platform Construction

After analyzing the MBSE design framework, the data exchange technology between the system modelling tool and the other engineering tools. It can be tried to build a possible MBSE design platform using mature engineering tools.

IBM DOORS is a common requirement management software. On the one hand, the requirements stored in DOORS can be exported as ReqIF format files and imported into system modelling tools, such as MagicDraw or Rhapsody. On the other hand, requirements can also be synchronized between Doors and MagicDraw/Rhapsody based on the OSLC specification.

For SoS and Business modelling, UPDM or UAF is used as the model language, and system modelling tools, like MagicDraw or Rhapsody support the model process.UAF/UPDM based models can be transformed into SysML-based models using the XMI specification.

CATIA can be used as a mechanical modelling tool, and SolidWorks Electrical can be used as an electrical modelling tool. API-based data exchange can transform system architecture and interface information from MagicDraw/Rhapsody to CATIA or SolidWorks Electrical to perform detailed design.

SCADE Architect is a lightweight system modelling software and can be used as a bridge for MagicDraw/Rhapsody and SCADE Suite to perform software modelling. The SysML model can be transformed into the UML model based on the XMI specification.

Simfia or Medini is a system safety and reliability analysis software, SysML model, which can be transported from MagicDraw/Rhapsody to Simfia/Medini to perform model-based safety analysis(MBSA). Furthermore, on the other hand, MagicDraw or Rhapsody has its software plugin or language profile to perform MBSA.

Modelica is a multi-physics domain simulation language and has developed rapidly in recent years. And Dymola and SimulationX are the mainstream Modelica modelling tool. The SysPhs and SysML-Modelica Transformation Specification support the transformation of the SysML model to the Modelica model.

FMI is a widely adopted specification to support co-simulation. The model built in Matlab or AMEsim can be exported as FMU and imported into MagicDraw or Rhapsody to perform system parameter simulation.

Though the existing data-transform specifications support most of the software, some classic or self-developed software can not be linked. So multidisciplinary optimization tool-based, API-based, and data bus-based data exchange methods can be used to build a whole and perfect MBSE design platform.

In conclusion, a unified MBSE design platform framework is shown in Fig. 12.

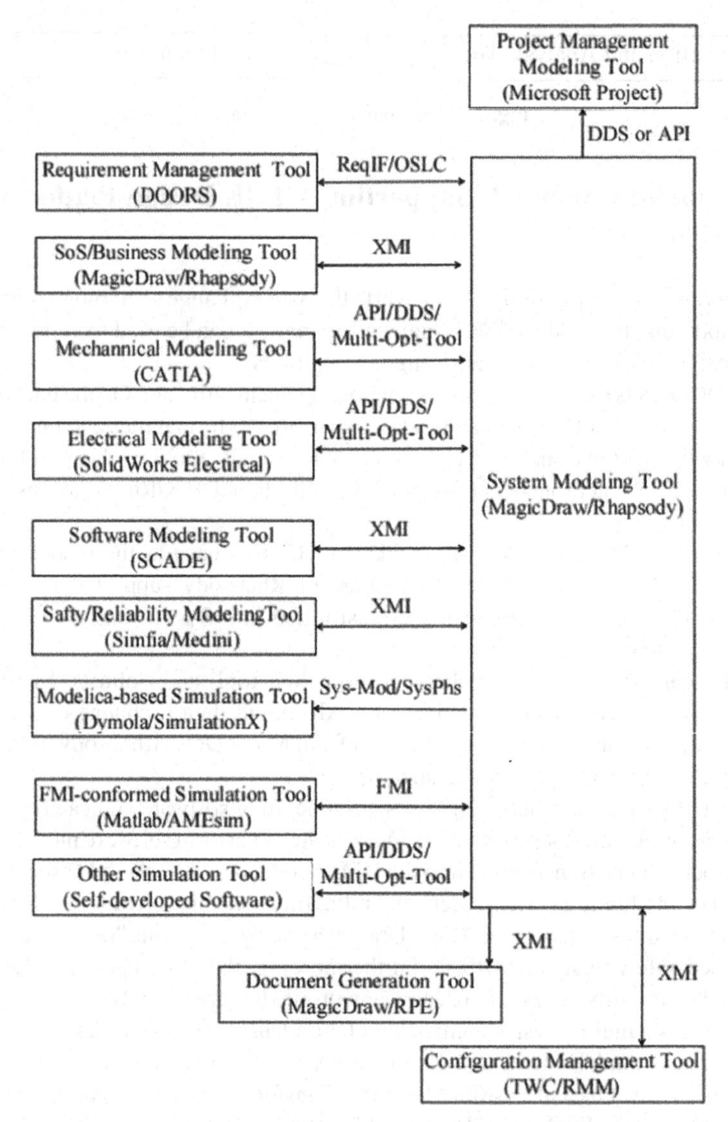

Fig. 12. A unified MBSE design platform framework

Main tools used in the platform framework are introduced in Table 1.

Table 1. Main tools used in the platform framework

Tool	Classification	Introduction
MagicDraw	System modelling tool	MagicDraw is a system modelling tool developed by NoMagic and later acquired by Dassault and integrated into the Dassault 3DE platform
Rhapsody	System modelling tool	Rhapsody is part of the IBM Engineering portfolio that provides a system modelling environment
DOORS	Requirements management tool	IBM DOORS is a common requirement management software
CATIA	Mechanical modelling tool	CATIA is the product design software developed and created by Dassault Systemes
SCADE	Software modelling tool	Ansys SCADE has an Embedded Software family. By using SCADE, system and software engineers can work within the same framework
Simfia	Safety and reliability modelling tool	APSYS Simfia is used to analyse and simulate its overall behaviour and automate R.A.M.S. studies
Medini	Safety and reliability modelling tool	Ansys medini analysis streamlines functional safety analysis across the entire system architecture
Dymola	Modelica modelling tool	Dassault Dymola is a complete tool for modelling and simulating integrated and complex systems using its Modelica and simulation technology
SimulationX	Modelica modelling tool	SimulationX is a sophisticated platform to efficiently model, simulate, and analyze mechanical, hydraulic, pneumatic, electrical, and combined systems
Matlab	FMI-conformed simulation tool	MATLAB is a programming and numeric computing platform to analyze data, develop algorithms, and create models
AMEsim	FMI-conformed simulation tool	Simcenter Amesim is a system simulation platform, allowing system simulation engineers to virtually assess and optimize the performance of mechatronic systems

(*continued*)

Table 1. (*continued*)

Tool	Classification	Introduction
RPE	Document generation tool	IBM Rational Publishing Engine is an automated document generation solution that generates documents from Rational products and selected applications from other vendors that use XML and REST interfaces
TWC	System model management tool	Teamwork Cloud is CATIA No Magic's next-generation repository for collaborative development and version model storage
RMM	System model management tool	Rhapsody Model Manager is a web-based application that integrates with Rational Rhapsody and other tools to provide lifecycle traceability for models

4 Case Study

MBSE can be used in civil helicopter fuel systems development to perform a virtual verification in the conceptual design phase to improve the design efficiency and quality. This needs a joint model and simulation platform that can be a suitable case to demonstrate the feasibility of building a unified MBSE modelling platform.

In this case, IBM DOORS is used to store and manage stakeholder requirements, and the requirements can be transferred to MagicDraw via the ReqIF file. In addition, requirements can also be synchronized between Doors and MagicDraw based on the OSLC specification (Fig. 13).

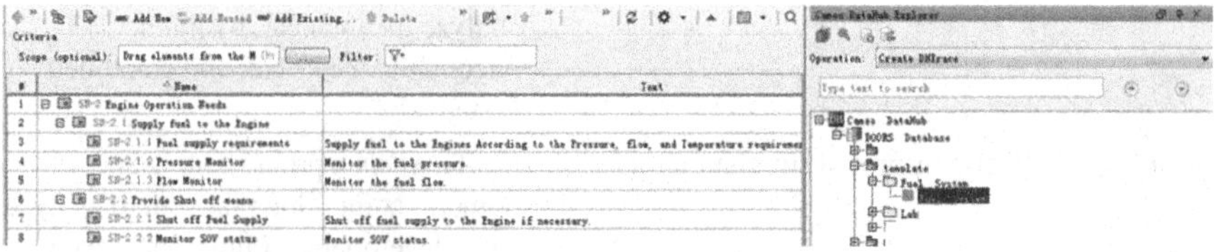

Fig. 13. Synchronize requirements between Doors and MagicDraw based on the OSLC

After the stakeholder requirements are imported, system context capture and use case modelling can be carried out. The SysML context block can be established in MagicDraw using the generation relationship to inherit as many SoS level modeling results as possible. The conversion of UPDM models to SysML models is based on the fact that they are both UML profiles, behind which the conversion between languages is performed through the XMI format (Fig. 14).

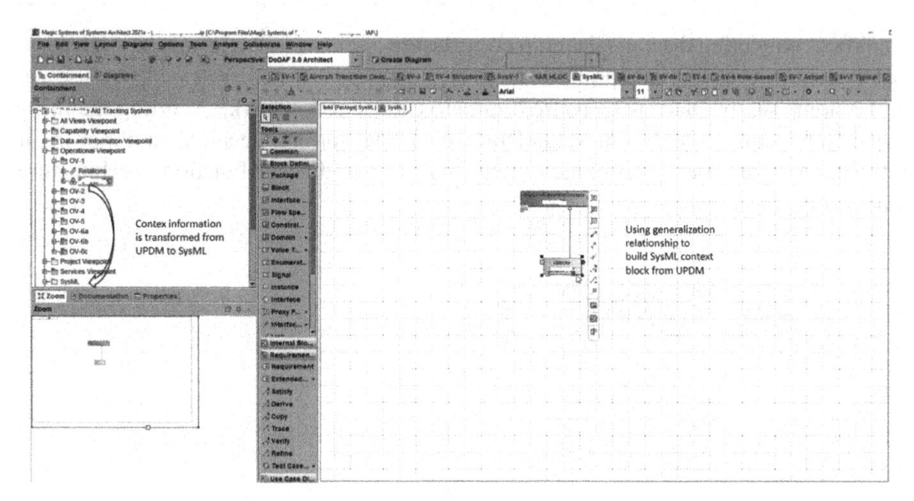

Fig. 14. Build SysML system context block from UPDM based on XMI

And to verify the fuel system MOEs, an AMESim fuel system simulation model is built and can be transformed into an FMU based on the FMI specification. The FMU can be imported into MagicDraw to Co-simulate with system model elements to perform the system verification (Fig. 15).

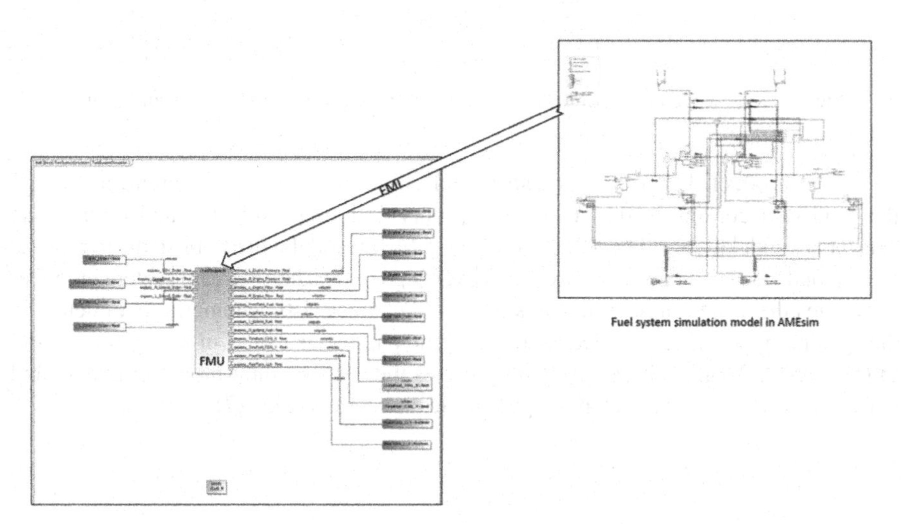

Fig. 15. Co-simulation of FMU and system model in MagicDraw

After completing the logical architecture design of the fuel system and the simulation verification of the MOEs, a preliminary safety analysis can be performed. We use the professional safety analysis software Simfia to carry out this work, exporting the functional architecture model from MagicDraw to an XMI format file and then importing it into Simfia to carry out a preliminary safety analysis based on the functional architecture model (Fig. 16).

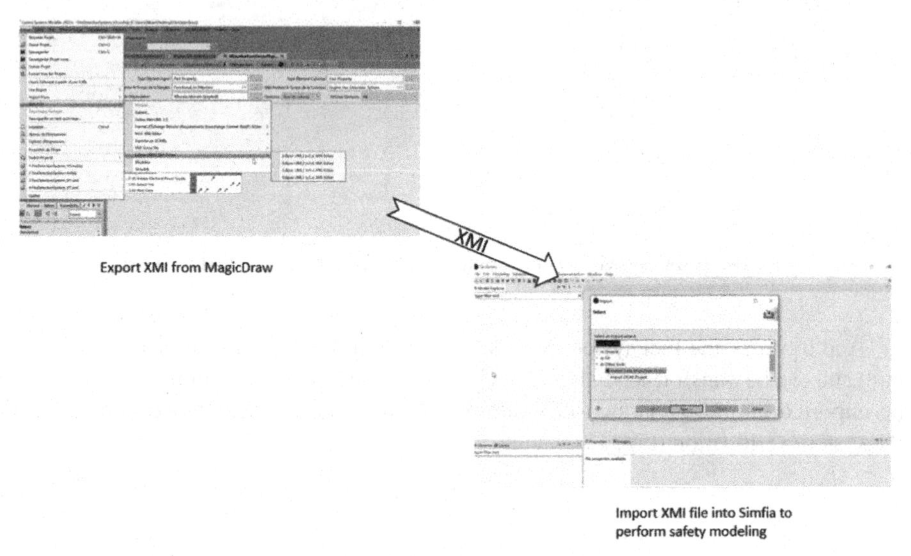
XMI
Export XMI from MagicDraw
Import XMI file into Simfia to perform safety modeling

Fig. 16. Exchange of functional architecture model between MagicDraw and Simfia

In the product design process, structural engineers and system engineers design the product according to the enterprise product design specification and refer to the enterprise model library. The MBSE model and MBD model will establish the parameter correspondence based on API-based data exchange. After the upstream system engineer completes the system model, it sends down the characteristic geometric parameters in the system model. It automatically transfers the parameter values in the system model to the CATIA 3D model, which directly drives the 3D structural model to update and realize the transfer of characteristic geometric parameters (Fig. 17).

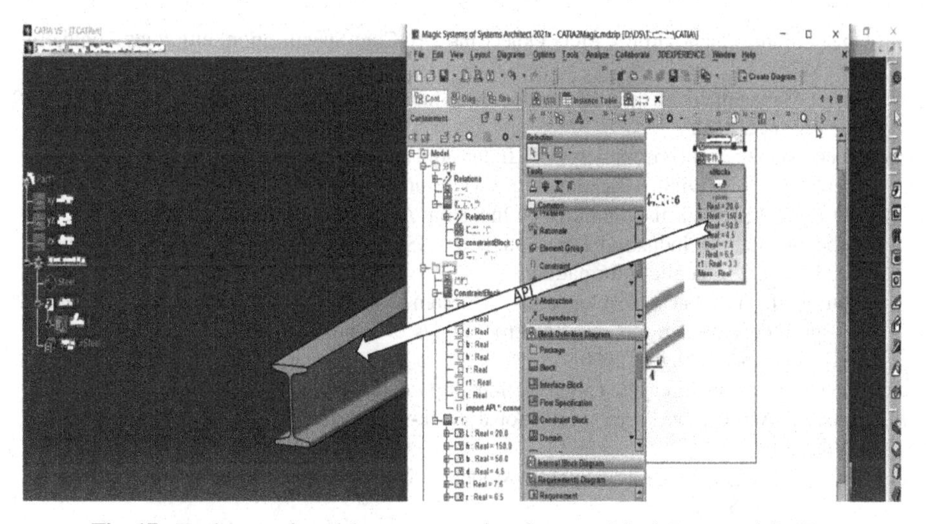

Fig. 17. Exchange of model parameter values between MagicDraw and CATIA

5 Conclusion

The system model built in the MBSE process can drive system safety/reliability analysis, system simulation, and downstream mechanical/electrical/software design. Lack of guidance to connect system modelling and other engineering tools leads to failure to take advantage of MBSE.

In this article, the MBSE design framework and the data exchanges between the tools are analyzed firstly. The four data exchange mechanisms are studied: modelling standard-based exchange, API-based exchange, multidisciplinary optimization tool-based exchange, and data bus-based exchange.

A unified MBSE design platform framework is given based on the mainstream engineering tools. In this framework, the system modelling tool is built in the centre of the platform like a hub. Compared with other platform frameworks like developing a platform based on API-based development, this solution fully uses the existing tools and data exchange specifications to make significant financial and time-saving.

In the end, a helicopter fuel system model as a case study demonstrates the feasibility of a unified MBSE modelling platform.

References

1. Lu, J.: An investigation of functionalities of future tool-chain for aerospace industry. In: INCOSE International Symposium, vol. 27, pp. 1408–1422 (2017)
2. Lu, J.: Towards a service-oriented framework for MBSE tool-chain development. In: 13th Annual Conference on System of Systems Engineering (SoSE), pp. 568–575. IEEE, Paris (2018)
3. Friedenthal, S.: A Practical Guide to SysML: The Systems Modeling Language, 3rd edn. Elsevier, Amsterdam (2015)

4. Object Management Group: Meta Object Facility Core Specification. http://www.omg.org/spec/MOF/
5. Object Management Group: XML Metadata Interchange (XMI) Specification. http://www.omg.org/spec/XMI/
6. Object Management Group: SysML Extension for Physical Interaction and Signal Flow Simulation (SysPhs) Specification. http://www.omg.org/spec/SysPhs/
7. ISO TC-184 (Technical Committee on Industrial Automation Systems and Integration). SC4 (Subcommittee on Industrial Data Standards). ISO 10303-233 STEP AP233. http://www.ap233.org/ap233-public-information/
8. Johnson, J.: The latest developments in design data exchange: towards fully integrated aerospace design environments. In: Proceedings of 22nd International Congress of Aeronautical Sciences, August 2000
9. Open Services for Lifecycle Collaboration. http://open-services.net/
10. Modelica Association Project: Functional mock-up interface (FMI). http://www.modelica.org/projects/
11. Object Management Group: SysML Modelica Transformation Specification. http://www.omg.org/spec/SyM/
12. Object Management Group: Requirements Interchange Format Specification. http://www.omg.org/reqif/
13. Pardo-Castellote, G.: Applying MBSE to the Industrial IoT: using SysML with Connext DDS and Simulink. In: NoMagic World Symposium 2018 (2018)

3D Visualization Supporting Situational Awareness of Model-Based System of Systems

Yuqian Cao[1], Boyu Xie[1], Jian Wang[1], Xiaochen Zheng[2], and Jinzhi Lu[2(✉)]

[1] University of Electronic Science and Technology of China, Chengdu, China
[2] Swiss Federal Institute of Technology Lausanne, Lausanne, Switzerland
jinzhi.lu@epfl.ch

Abstract. Model-based System of Systems Engineering (MBSoSE) enables to improve the efficiency of the architecture design, and the performance analysis and modeling of highly complex system of systems (SoS). Introducing situational awareness into complex system of systems engineering (SoSE) can facilitate users to understand and analyze the SoS models and make decisions through the situational awareness. However, conventional texts or models cannot be intuitively understood by users through traditional SoSE approach. Therefore, this paper proposes a 3D visualization technology to support situational awareness using MBSoSE. A semantic modeling solution based on KARMA language and Graph-Object-Port-Property-Relationship-Role (GOPPRR) ontology is proposed to combine 3D visualization technology with co-simulation to enhance situational awareness through enhanced vision. It can enhance the capabilities of perception, understanding and prediction of decision makers. A case study is conducted to prove the effectiveness of the proposed solution which the braking scene is modeled in an intelligent vehicle SoS using the 3D visualization technology to dynamically co-simulate the braking scene.

Keywords: MBSE · KARMA · 3D visualization · Situational awareness

1 Introduction

System of systems (SoS) is a complex engineering system composed of many engineering systems [1]. With the development of human society, a variety of systems with complex structures, diverse functions and large scales are proposed. Different systems can provide functions that cannot be realized by a single system through cooperation [2].

Traditional System of Systems Engineering (SoSE) is facing several challenges. First of all, a SoS often involves multiple engineering disciplines, several development teams, and different stakeholders. Various domain specific architecture description languages are used to describe the SoS. It may lead to misunderstandings among developers which can further result in unified SoS models. Secondly, the traditional MBSoS is usually implemented based on the life cycle model [3], according to which the SoS model design is carried out by models first and then the production of the physical prototype is

J. Chen et al. (Eds.): KSS 2022, CCIS 1592, pp. 113–127, 2022.
https://doi.org/10.1007/978-981-19-3610-4_8

carried out. This approach may result in the gap between the real products and SoS design because less verification is implemented before prototyping. The iterative development process led by inefficient verification may greatly increase the production cost and time cycle. Finally, decision-support through modeling and simulation is one of the critical tasks for SoS design. Traditional SoSE requires decision makings for different system capability analysis. Technologies, such as situational awareness has been widely used in order to minimize uncertainty, production costs, and eliminate barriers for verifying the SoS models [4].

In the field of MBSoS, situational awareness is a key factor that determines the effectiveness and timeliness of the system which consists of three independent layers: perception layer, understanding layer and prediction layer. The perception layer mainly senses the state, attributes and dynamics of related elements in the environment. The understanding layer is based on the perception layer to understand the meaning of data and clues to the target. The prediction layer predicts the ability of elements in the short term based on the perception layer and understanding layer.

3D visualization is considered as one important technique to present complex and abstract data information with appropriate visual elements and perspectives, which is easy for users to understand, remember and transfer the SoS development information for supporting situational awareness [7]. Combining visualization with situational awareness can solve the problems mentioned above in MBSoSE [5].

- First of all, compared with other formats of situational awareness, visual vision contains the largest amount of dynamic information during SoS operations. 3D Visualization is used to enhance situational awareness, by collecting more information at the perception layer in order to provide support for the understanding layer and prediction layer.
- Second, the introduction of the 3D visualization in the understanding layer enables the system developers to identify and to understand the rational between the environment factors, and the development trend of the elements. It minimizes the understanding deviations of a model caused by inconsistent knowledge of cross-domain developers.
- Finally, the combination of situational awareness and 3D visualization can enhance the functional capabilities of the situational awareness prediction layer [6].

  - On one hand, it can visualize the system design process through the knowledge required for verification. Therefore, the development and design can be constantly adjusted through a more understandable visualization during the verification process. Correspondingly, the production cost and cycle time can be properly controlled with less risks lead by the iterative design.
  - On the other hand, after the integrating MBSoSE, situational awareness can be used in a specific scene to understand the dynamics of each system in a certain time and space.

In this paper, 3D visualization is applied to complex SoSE situation awareness. The data of visualization file is converted into CZML format for static display with 3D models in a developed plugin. Meanwhile, With co-simulation, the situation awareness is implemented with simulations implemented within the SoS model development. The scene is created based on a 3D visualization model combining the hybrid automata simulation execution. It can help SoS designers realize the integration of the dynamic performances of SoS operations, thus to reduce the complicated SoS development cost and time. At the same time, it is beneficial for the users to make quick decisions in specific scenarios.

The rest of the paper is organized as follows. In the second chapter, related work is discussed, including the existing problems in SoSE, the proposal and development of situational awareness in the sSoSE, and how the current technology realizes SoS modeling and situational awareness. Then, the method of this paper is introduced in Sect. 3. SoS modeling is carried out on MetaGraph 2.0[1] using KARMA language based meta-model library and model for SoS. 3D visualization is integrated into MetaGraph 2.0 for the situational awareness, based on the 3D visualization models which are configured to implement co-simulation for 3D visualization dynamic simulation for specific scenes. In Sect. 4, a case study of auto-braking scene in intelligent vehicle SoS is presented, and the role of the proposed method during automatic driving system process is discussed and analyzed. Finally, the conclusion is given in Sect. 5.

2 Related Work

SoSE mainly experienced three stages: text-based SOSE, MBSoSE and simulation modeling SOSE. There are many problems in text-based SOSE, such as too much text and not easy to understand. Daniele Gianni et al. proposed an ESA Architecture Framework (ESA-AF) in 2012 to support SoSE activities such as Galileo Navigation, Global Environment and Safety Monitoring (GMES) and Space Situational Awareness (SSA) projects [8]. However, MBSoSE using the architecture approach still has problems, such as the inconsistency between the product and the design in the physical integration stage, which increases the cost and production cycle [9]. Thus, Zhang Lin et al. proposed Modeling and Simulation Based System of Systems Engineering (MSBS2E) [1], aiming at modeling the full life cycle of products, digital integration of products through simulation technology for verification, testing and situational awareness of specific scenarios in complex SoS [10].

Situational awareness was first proposed by the US Air Force in the 1980s and used in aerospace research. It mainly consists of three levels: perception, understanding and prediction [11]. In Wikipedia says *"Situational awareness (SA) is the perception of environmental factors and events related to time or space, the understanding of their meaning, and the prediction of their future state"* [12]. In 1988, Endsley proposed that situational awareness is a process of acquiring various elements in the environment from the perspective of time and space, understanding these elements and predicting their future state. Using situational awareness was conducive to eliminating the barriers between different professionals in complex systems [13]. In 1995, Endsley proposed that situational awareness was important for system operation. He proposed a situational awareness model based on human dynamic decision making, which can be used in specific scenarios of complex systems, using the predictive function of situational awareness to support system decision making [14].

[1] http://www.zkhoneycomb.com/

In complex SoSE, situational awareness undertakes the functions of removing domain barriers in the design and development stage, supporting simulation verification, and supporting user decision-making through situational awareness in the application stage. The above functions are mainly realized by combining the visualization process of complex system model with situational awareness.

In 1999, Card put forward the information visualization model, which mainly includes original data, data table, visual structure and view [15]. In addition, raw data, data tables, visual structures and views could be divided into data layer, visual view and view interaction to support the visualization process in situational awareness. The three layers of situational awareness are respectively supported by different data in the data layer. The overview view, the detail view, and the coexistence of multiple views in the visual view can be used in the decision stage [16].

Based on the above research, it can be found that situational awareness can help eliminate domain barriers, support simulation verification, and support user decision-making in complex systems. 3D visualization technology can enhance situational awareness by enhancing vision. Therefore, this paper proposes a method that 3D visualization supports situational awareness of MBSoS. This paper mainly adopts a semantic modeling approach based on KARMA language and GOPPRR ontology [17] in Metagraph 2.0 platform to realize system meta-model development and system modeling and a 3D visualization approach for situation awareness. First, KARMA language is used to develop SoS models based on UPDM meta-models [18].Then system behaviors are defined within the KARMA models in order to support dynamic simulation of the SoS operation scene. During the system simulations, the data of visualization file is first developed and converted into CZML format data for the static display of 3D model in the 3D visualization plugin. At the same time, the variables in the KARMA model are define to map to the corresponding variables in CZML data used for 3D visualization models, so that the system design and situational awareness based on 3D visual model are integrated to implement co-simulation.

The advantages of the proposed method in this paper are twofold:

1. This paper adopts semantic a modeling approach based on KARMA language and GOPPRR ontology to realize the development of SoS meta-model, SoS modeling, SoS simulation and visualization.
2. 3D visualization is used to be integrated into Metagraph platform and realize situation awareness of complex SoSE by configuring 3D visualization model for simulations.

3 3D Visualization Model Construction for MBSoSE Situational Awareness

3.1 The Overall Framework

As shown in Fig. 1, an integrated framework based on KARMA language and 3D visual simulation for MBSoS situation awareness is proposed. There are three main steps to implement the entire 3D simulation and visualization process:

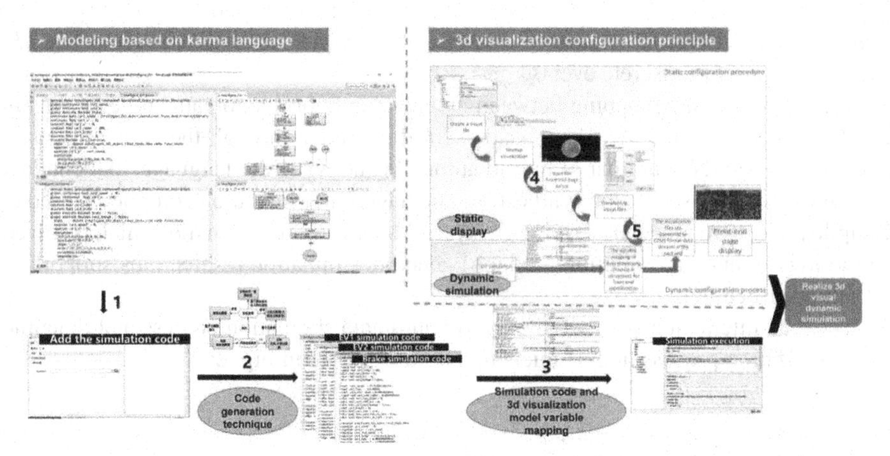

Fig. 1. An integrated framework based on KARMA language and 3D visual simulation for MBSoS situation awareness.

1. Develop SoS models based on KARMA UPDM meta-models. In order to define the SoS formalism, meta-models including 56 graphs are used to define different architectural views of SoS.
2. When developing the brake models for system status definition, variables and equations based on hybrid automata are defined based on KARMA language.
3. After developing the KARMA models, compiling for co-simulation is executed to transform the KARMA models representing system status into execution codes.

   After the preparing the co-simulation execution codes, several steps are implemented to develop 3D visualization models and execute the simulation.
4. 3D visual file is developed with 3D models in order to construct the situation awareness scene as shown in Sect. 3.3.
5. The 3D visual models are transformed into CZML codes in order to demonstrate the 3D earth scene.

In order to support the co-simulation between KARMA models and 3D visualization models, a co-simulation configuration is implemented to map the variables generated from hybrid automata simulation to the variables in the 3D visualization scene. Finally, a co-simulation is implemented.

3.2 Co-simulation for 3D Visualization and Hybrid Automata Simulation

We use hybrid automata to describe the behaviors in the system, and the hybrid automata model can be represented by a mathematical model, as shown in Eq. 1:

$$HA = (DSS, TF, FS, VIS, INV, FL, JU, DE) \tag{1}$$

The mathematical model is an 8-tuple. *DSS* presents a limited set of discrete states.*TF* presents transitions between discrete states. *FS* presents a limited set of variables. *VIS* presents a set of initial states. *INV* presents a set of invariants. *FL* presents the continuous evolution in one state. *JU* presents a set of functions which allocated to each of the *TF*. *DE* presents a set of discrete events.

We set up specific mapping between elements in hybrid automata and elements in KARMA model, define the grammar of hybrid automata through the above mapping rules, synthesize the grammar of hybrid automata and KARMA language, and conduct system modeling based on the synthesized language. The simulation compiler is used to compile the system model describing the system behavior into a.sim file, which is solved by the hybrid automata solution model to generate simulation results, thus supporting dynamic verification.

The data interaction between the hybrid automata simulation process and dynamic 3D visualization process in dynamic simulation is shown in Fig. 2:

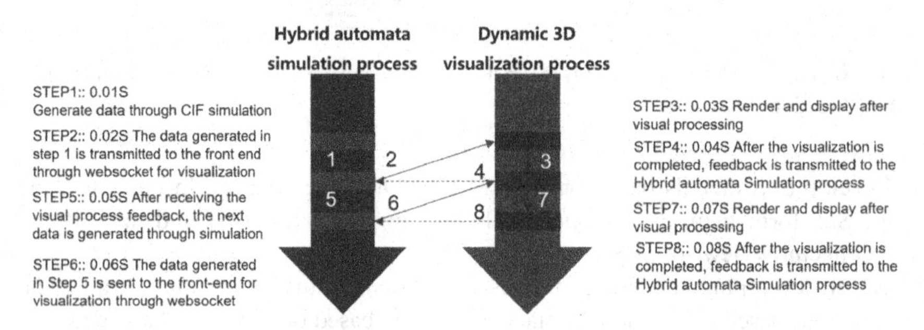

Fig. 2. Hybrid automata simulation process and dynamic 3D visualization process.

The steps are as follows:

1. Generate data through CIF simulation in the hybrid automata simulation process.
2. Transmit the data generated in step 1 to the front end through websocket for visualization.
3. Render and display after visual processing in the hybrid automata simulation process.
4. Transmit feedback to the hybrid automata simulation process after the visualization is completed.
5. Generate the next data through simulation in the hybrid automata simulation process after receiving the visual process feedback,
6. Repeat the process until the end of the simulation.

The co-simulation solution mechanism in the hybrid automata simulation process is described below. The co-simulation system can be regarded as a discrete system composed of several subsystems. In each subsystem, the state variables are shown in Eq. 2, the output equation is shown in Eq. 3. The input and output of each subsystem exchange data at a specified point in time. We calculate the input vector of each subsystem and the output of other subsystems by a coupling formula, as shown in Eq. 4. This coupling formula connects the different subsystems in the co-simulation.

$$x^i = k^i(x^i, u^i, t), x^i(t_0) = x^i_o \tag{2}$$

$$y^i = h^i(x^i, u^i, t), i = I, II, III, N \tag{3}$$

$$u^i = M^i y = [M^{i,I}, \ldots, M^{i,i-1}, 0, M^{i,i+1} \ldots M^{i,N}] \begin{bmatrix} y^I \\ . \\ . \\ y^{i-1} \\ y^i \\ y^{i+1} \\ . \\ . \\ y^N \end{bmatrix} \tag{4}$$

i stands for different subsystems, $x^i \in R^{n^i_x}$ is the state variable, $x^i \in R^{n^i_u}$ is the input vector, $y^i \in R^{n^i_y}$ is the output vector, $t \in R$ stands for time, i stands for subsystem, $M^{i,j} \in R^{n^i_x * n^i_y}$ stands for the matrix element, and can be 0 or 1.

3.3 3D Visualization Model Configuration

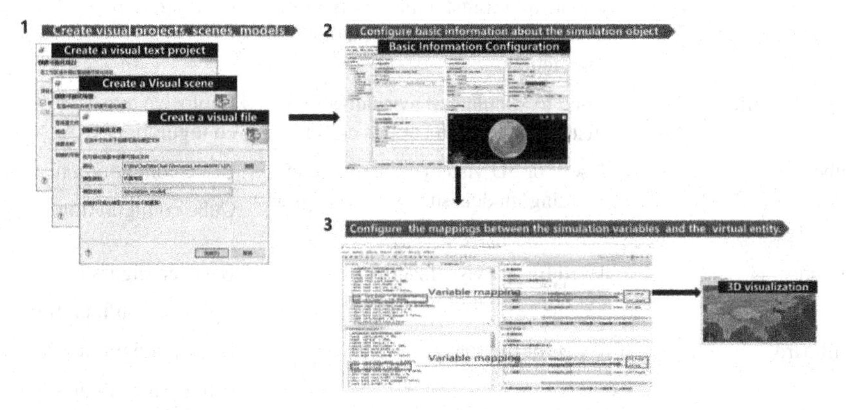

Fig. 3. 3D visualization model configuration process.

This section introduces building a 3D dynamic visualization for a co-simulation code file based on Sect. 3.2. In order to construct 3D dynamic visualization, 3D visual models are developed in a 3D visualization plugin in MetaGraph 2.0 as shown in Fig. 3. The specific steps are as follows:

1. Create a visual scene to represent the visual earth scene.
2. Add static entities as shown in Table 1 in order to construct all the 3D entities in the visualization
3. Configure the mappings between the simulation variables which are generated in each step by hybrid automata simulation solver and the status of each virtual entity.

Table 1. Visual entity for developing visualization scene.

Visual model	Meaning	Basic information
The simulation model	3D visualization models such as car, radar, passenger plane, fighter plane, control center, satellite and warship can be configured for the ontology. The 3D visualization model is associated with the simulation through variable mapping	Emulation object basic configuration
		Billboard configuration
		The tag
		The path configuration
		Location configuration
		Variables mapping
CZML header file	A visual simulation model for connecting transferable information as a visualization of information interaction	CZML Basic information
Radar wave	A visual simulation model for simulating radar scanning waves that can detect objects in the scanning range	Radar scan effect configuration
Information line	A visual simulation model for connecting transferable information as a visualization of information interaction	Information line configuration
Explosion effect	Used to simulate 3D visualizations that explode when entities collide	Explosive effect configuration
Cube	Used for 3D visualization of cubic building models such as buildings and warehouses	Basic configuration
		Cube configuration
Cylinder	Used for 3D visualization of cylindrical buildings or models	Basic configuration
		Cylinder configuration
Billboard	Used for billboard 3D visualization	Basic configuration
		Billboard configuration
Cone	Used for 3D visualization of cone models	Basic configuration

(*continued*)

Table 1. (*continued*)

Visual model	Meaning	Basic information
		Cone configuration
Polyhedron	Used for 3D visualization of polyhedral objects	Basic configuration
		Polyhedron configuration
Sphere	Used for 3D visualization of ball-shaped objects	Basic configuration
		Sphere configuration
Line	Used for 3D visualization of linear objects	Basic configuration
		Line configuration
Dot	Used for 3D visualization of point objects	Basic configuration
		Dot configuration

The 3D visualization model in Table 1 includes dynamic model and static model. The dynamic model means that the model state will change over time, and the model state variables need to be bound with the variables in the simulation code to carry out dynamic simulation, including simulation model, radar wave, information line and explosion effect. The model state of a static model remains constant during the simulation process and is usually used to simulate the environment of a simulation scenario. The static model includes cube, cylinder, billboard, cone, polyhedron, sphere, line and dot.

4. After the basic information of the visualization model is configured, we can transfer the CZML files to the visualization process or enter the 3D display page by dragging and dropping the simulation file.

Therefore, the 3D visualization configuration has the following advantages. On the one hand, by configuring the 13 visualization models shown in Table 1, entities can be simulated realistically. On the other hand, the configuration of visual model and the writing of hybrid automata simulation code can be carried out simultaneously, and finally associated the two processes by variable mapping, so that we can improve the efficiency of 3D visualization model configuration.

4 Case Study

In this case, KARMA meta-model library and models are used for SoS modeling in MetaGraph 2.0. 3D visual simulation is used to visualize the auto-braking scene which is executed through co-simulation from KARMA models. In this chapter, the problem statement is first introduced, then KARMA models to represent SoS formalism. Finally, configuration of 3D visualization and co-simulation are introduced to support the dynamic situation awareness of the auto-braking.

4.1 Background

In the case study, two vehicles are moving to the same direction on the road as shown in Fig. 4A. When the *car_1* radar detects ahead road barriers, *car_1* will brake for deceleration; at the same time, when the radar detector of *car_2* detects that the distance from *car_1* is less than a certain value, *car_2* will also brake according to the moving state of *car_1*. At the same time, information is transmitted between vehicles, satellites, signal towers and radars.

According to the auto-braking scene, the entire workflow includes, a total of nine operational tasks as shown in Fig. 4B. Firstly, *car_1* radar scanning is open; if there are obstacles ahead, *car_1* will receive the reflected radar signal of the obstacles, and *car_1* request GPS signal; after *car_1* received GPS signals, it sends radar signal and GPS information to *car_2*; and *car_1* confirms that *car_2* has received the information by confirming the radar signal sent by *car_2*. On the other hand, *car_1* sends a signal to the control center, which then sends a message to *car_1* after the control center processed the messages. After receiving the information from *car_2* and the control center, *car_1* will start braking to complete the transformation of the entire combat state.

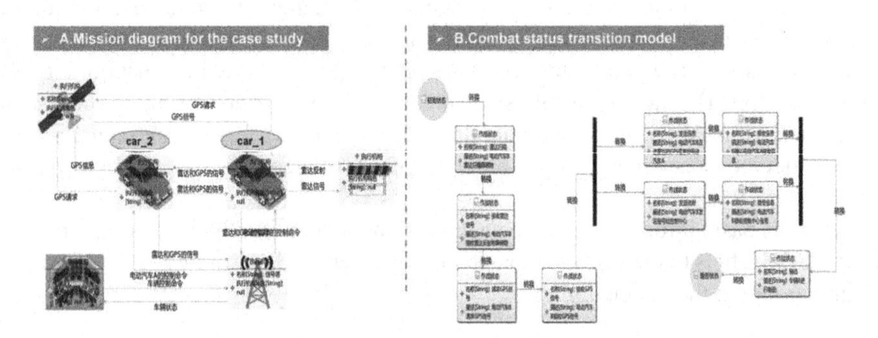

Fig. 4. Auto-braking scene model description

4.2 The SoS Models for System Behaviors Based on Hybrid Automata Simulation

In order to define the SoS operations and system capabilities, SoS models are defined as shown in the Fig. 5: the design operational analysis model, state transition model, and etc. Figure 5C is KAMAR language corresponding to the state transition model, Fig. 5D is a .sim file compiled from the state transition model.

- The design operation analysis model is mainly used to define the use case related to the SoS scenario. It includes three key stakeholders and their system concerns in Fig. 5A.
- The state transition model describes the vehicle's motion state in order to support the auto-braking operation in Fig. 5B. System behaviors are defined based on KARMA language, as shown in Fig. 5C.

- Figure 5C shows the KARMA language corresponding to the state transition model. The KARMA language is based on GOPPRR ontology, and the KARMA model is an instance of the graph meta-model, which includes object meta-model, point meta-model, property meta-model, relationship meta-model and role meta-model.
- Figure 5D shows the .sim file of the state transition model compiled by the simulation compiler. In the figure, *automaton* represents the state machine of the state transition model. Each state is marked with *location* in the .sim file, *initial* represents the initial state, *edge* represents the state jump, *when* represents the condition to be met for the state jump, and *goto* indicates the next state of the jump.

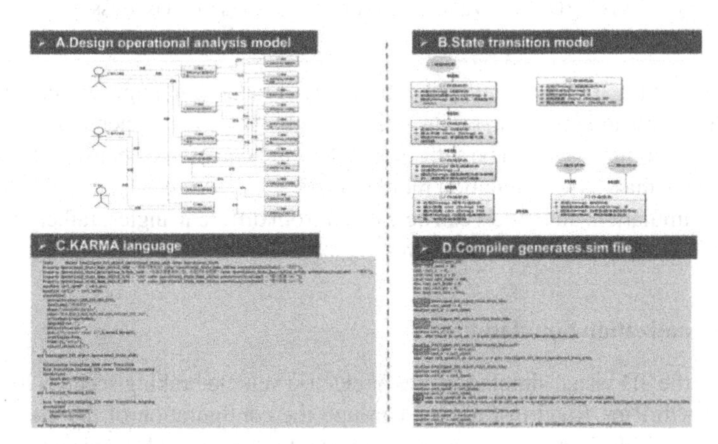

Fig. 5. The SoS models, KARMA language and .sim file

4.3 Configuration for 3D Model Development and Co-simulation

In order to develop 3D models and to configure co-simulation between KARMA models and 3D visualization models, the configuration process is shown in Fig. 6.

When developing the 3D models, visual entities as shown in Table 1 are developed in order to construct the entire situation awareness scenario. 3D visualization model can be divided into dynamic model and static model. Dynamic model includes satellite, control center, vehicle, radar scanning wave and information line. In addition to the configuration of basic information such as size, color, name and initial position, variables of the dynamic model should be bound with corresponding variables in the co-simulation code to support 3D visual dynamic simulation. Information lines should be configured between vehicles and satellites and between vehicles and control center for information interaction. The detailed information interaction process is described in Sect. 4.1. The static model includes cubes and other geometry used to simulate buildings in the braking scene, and its 3D visualization configuration mainly includes size, initial position, color and so on.

Compared with traditional tables or 2D images, the advantages of 3D visualization model in this case are highlighted in the following aspects:

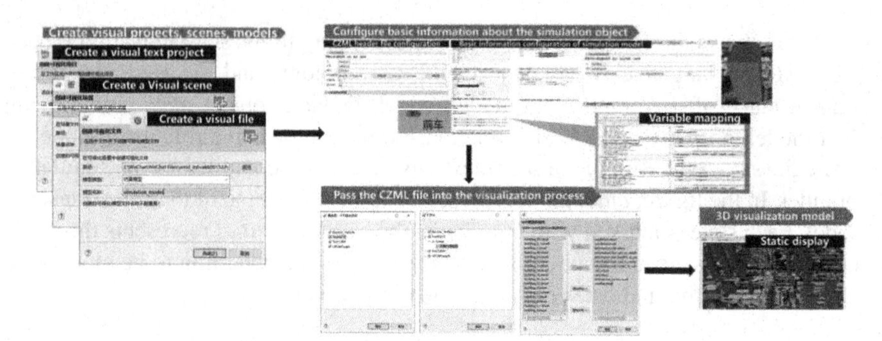

Fig. 6. 3D visualization simulation of configuration process

1. Compared with 2D simulation, only the radius of radar scanning wave can be set. In 3D simulation, the direction, radius, left and right Angle and upper and lower Angle of radar scanning wave can be set, so as to determine a three-dimensional scannable range. This makes the simulation more realistic.
2. During simulation, all models can be viewed from different angles, rather than being limited in a 2D plane, which improves situational awareness.

4.4 3D Visualization Simulation

In this case, the 3D visual simulation site of electric vehicle braking scene is located on the East Fourth Ring Road in Beijing. Through the configuration of the visual model in Sect. 4.3, variable mapping is used to associate the variables of the visual simulation model with the corresponding variables in the simulation code, so as to achieve the purpose of simulation and visual linkage. The 3D visual dynamic simulation of the car braking scene is shown in the video (https://www.bilibili.com/video/BV19L411A7HL?spm_id_from=333.999.0.0).The visual configuration model of this case is shown in Table 2:

Table 2. Auto-braking scene visualization model statistics.

Visual model	Type	Number
The simulation model	The car	2
	Satellite	1
	The control center	1
Information line	–	4
Radar wave	–	1
The cube	–	90

After the above model configuration is completed, the model is transferred to the visualization program for simulation execution. The simulation results and auto-braking scene 3D visualization simulation is shown in the Fig. 7 below:

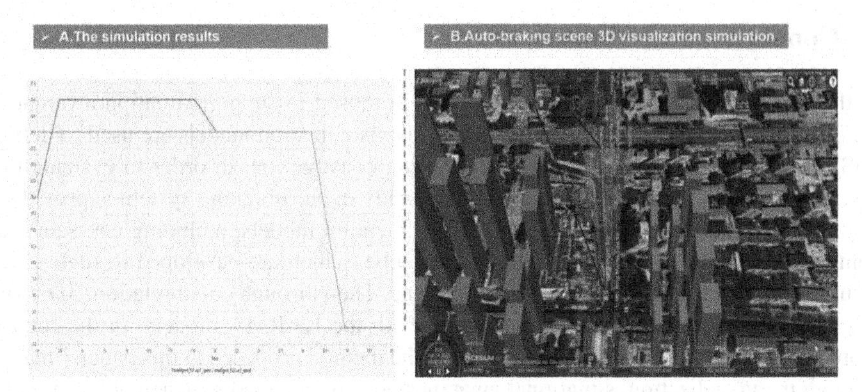

Fig. 7. The simulation results and auto-braking scene 3D visualization simulation

Figure 7A is the speed-time image of two cars, As can be seen from the image, when the speed of *car_2* increases to a certain value, it begins to move at a uniform speed. When the distance between *car_1* and *car_2* is less than a certain value, it slows down to the same speed with car 1. When the distance between car 1 and obstacle is less than a certain value, the two cars slow down at the same time. And Fig. 7B is the auto-Braking scene 3D visualization. Compared to speed-time images, 3D models can quickly enable interdisciplinary experts to understand auto-braking scene; Compared with 2D simulation, 3D dynamic simulation can more restore the real space, enhance the authenticity and detail of the visual level of situation awareness, and help decision makers to make decisions.

4.5 Discussion

From the case study, we find six visualization models including car, satellite, control center, information line, radar wave, cube. They are developed to realize the co-simulation of the auto-braking scene. Through a 40-s simulation, technical developers can have the most intuitive understanding of the auto-braking process without deep professional knowledge. During the visualization, visual efforts, such as information lines and radar waves are integrated with transparent data transmission with physical entities, which are generated from co-simulation. The visualization enhances the visibility of data and knowledge extraction, data interaction and data processing in in order to support MBSoSE situational awareness, and enables users to have a deeper understanding of system dynamics during SoS operations.

3D visualization is used to support MBSoSE situational awareness in the auto-braking scene. It can display the relative position and shape proportion between different vehicle entities which received data from simulations based on KARMA models, which is convenient for users to receive information more intuitively, to understand the system dynamics, and verify the system capability based on the simulation results. Moreover, 3D visualization can enhance situational awareness from the perspective of enhanced vision, thus improving the accuracy of decision making by SoS developers.

5 Conclusions and Future Work

In this paper, a 3D visualization approach is proposed to support situation awareness for SoS development. KARMA models and 3D visualization models are used to define SoS formalism and simulation awareness scene construction. In order to evaluate the availability of the proposed approach, a case study of auto-braking system is provided. In order to construct the 3D models, six visualization models, including car, satellite, control center, information line, radar wave, cube, which are developed to realize the visual scene definition of the auto-braking scene. Then through co-simulation, 3D visualization models are executed synchronized with the KARMA models. In the future, more engineering cases will be used to verify the method proposed in this paper. Finally, through the visualization, situational awareness is implemented which enables users to have a deeper understanding of system dynamic during the SoS operations.

Acknowledgement. This work was supported by Embedded processor security scanning technology for physically isolated network (2021YFG0338) and Research on security chip for complex electromagnetic environment simulation system (2021YFG0008).

References

1. Lin, Z., Kunyu, W.: System engineering based on modeling simulation. J. Syst. Simul. **34**(2), 179 (2022)
2. Topper, J.S., Horner, N.C.: Model-based systems engineering in support of complex systems development. Johns Hopkins APL Tech. Digest **32**(1) (2013)
3. Zhang, L.: Applying system of systems engineering approach to build complex cyber physical systems. In: Selvaraj, H., Zydek, D., Chmaj, G. (eds.) Progress in Systems Engineering, pp. 621–628. Springer, Cham (2015). https://doi.org/10.1007/978-3-319-08422-0_88
4. Durso, F.T., Hackworth, C.A., Truitt, T.R., et al.: Situation awareness as a predictor of performance for en route air traffic controllers. Air Traffic Control Q. **6**(1), 1–20 (1998)
5. Austin, M., Mayank, V., Shmunis, N.: PaladinRM: graph-based visualization of requirements organized for team-based design. Syst. Eng. **9**(2), 129–145 (2006)
6. Herter, J., Ovtcharova, J.: A model based visualization framework for cross discipline collaboration in industry 4.0 scenarios. Procedia CIRP **57**, 398–403 (2016)
7. Ahram, T., Karwowski, W., Amaba, B., Fechtelkotter, P.: Power and energy management: a user-centered system-of-systems engineering approach. In: Yamamoto, S. (ed.) HIMI 2013. LNCS, vol. 8017, pp. 3–12. Springer, Heidelberg (2013). https://doi.org/10.1007/978-3-642-39215-3_1
8. Piaszczyk, C.: Model based systems engineering with department of defense architectural framework. Syst. Eng. **14**(3), 305–326 (2011)
9. Mažeika, D., Butleris, R.: MBSEsec: Model-based systems engineering method for creating secure systems. Appl. Sci. **10**(7), 2574 (2020)

10. Gianni, D., Ambrogio, A.D., Tolk, A.: Modeling and Simulation-Based Systems Engineering Handbook. Taylor & Francis Group, Boca Raton (2015)
11. Salmon, P., Stanton, N., Walker, G., et al.: Situation awareness measurement: a review of applicability for C4i environments. Appl. Ergon. **37**(2), 225–238 (2006)
12. 2019. https://en.wikipedia.org/wiki/Situation_awareness
13. Endsley, M.R.: Design and evaluation for situation awareness enhancement. In: Proceedings of the Human Factors Society Annual Meeting, vol. 32, no. 2, pp. 97–101. Sage Publications, Los Angeles (1988)
14. Endsley, M.R.: Toward a theory of situation awareness in dynamic systems. In: Situational Awareness, pp. 9–42. Routledge (2017)
15. Card, M.: Readings in Information Visualization: Using Vision to Think. Morgan Kaufmann, San Francisco (1999)
16. Wu, J., Wang, J.: Visual perception Model based on situation Awareness Theory. Modern Libr. Inf. Technol. **7**, 9–14 (2010)
17. Ding, J., Reniers, M., Lu, J., et al.: Integration of modeling and verification for system model based on KARMA language. In: Proceedings of the 18th ACM SIGPLAN International Workshop on Domain-Specific Modeling, pp. 41–50 (2021)
18. Lu, J., Wang, G., Ma, J., et al.: General modeling language to support model-based systems engineering formalisms (part 1). In: INCOSE International Symposium, vol. 30, no. 1, pp. 323–338 (2020)

HRSE: Reframing the Human Resource Management with MBSE

Yuejie Wen[1(✉)], Li Wen[1], Zhaocai Zhang[2], Shan Gao[1], and Yapeng Li[3]

[1] China Academy of Space Technology, Beijing 100094, China
wenyuejie@126.com
[2] Beijing Institute of Space Science and Technology Information, Beijing 100094, China
[3] Space Star Technology Co., Ltd., Beijing 100095, China

Abstract. According to the ISO/IEC/IEEE 15288 standard, human resource management process is an essential part of organizational project-enabling processes, becoming more and more important during the process of science, technological and industrial innovation. However, many enterprises get lost when they confront overlapped complexity such as confusion of terminology systems, adaption of interwoven development pipeline system, and optimization of workforce topology. Therefore, this article presents a model-based approach for human resource management in enterprises, providing improved cognition, analytics and implementation. In particular, a comprehensive MBSE toolbox and a set of models based on KARMA language are constructed for effective human resource management. Moreover, within the specific domain of human resource management, a system of systems is built up through a strategic initiative, guiding tens of thousands of employees and dozens of organizations towards long-term goals of high-quality development. The case study and measurement results demonstrate that human resource systems engineering (HRSE) conforms to the equivalent applicability of MBSE methods.

Keywords: Human resource management · MBSE · High-quality development · KARMA language

1 Introduction

Model-based systems engineering (MBSE) and model-based design/definition (MBD) methods have been applied to many complex systems/missions in aerospace and aeronautical domain. The effectiveness of system thinking, and applicability of related model-based technologies are evident and proven in product engineering. In a typical MBSE approach, much of the information is captured in a set of system models [1]. Many cases of the aerospace and aeronautical missions have witnessed the effectiveness of the system thinking and related technologies of this approach to improve the development process of products [2].

J. Chen et al. (Eds.): KSS 2022, CCIS 1592, pp. 128–138, 2022.
https://doi.org/10.1007/978-981-19-3610-4_9

As the founder of China's systems engineering theory and practice, Mr. Qian Xuesen summarized the systems engineering methods with Chinese characteristics based on the experience of chinese aerospace industry, and further developed the theories of system science architecture and meta-synthesis method [3]. Chinese space industrial engineers deeply embraced the system thinking and systems engineering so long time, that systems engineering has been elevated to cultural DNA or totem of many aeronautical companies, even become an indispensable part of national spiritual pedigree. Systems engineering method has penetrated into many affairs of technical management and enterprise engineering, far beyond the technical application scope.

Meanwhile, the social discipline including the management engineering subject to continuous change [4]. Enterprise long-term planning takes a strategic view of the major plans and processes needed for an organization to achieve its mission [5]. We live in such a period of profound transition, and things are changing fast in human resource management. Technologies such as - cloud computing, data mining and talent analytics etc. – elevate the human resource management, to a systematic and strategic view, such as formulating various plans for boosting employee performance and engagement, with more and more measurement indicators [6].

Human resources – especially the high-performance talents – become more and more important. Each talent is an extremely complex modeling object, including various explicit attributes like age, education background, work experience, performance records, social network, intellectual properties, etc., and implicit profile like character, leadership, capability, expectations, ambitions etc. Workforce of an organization may have many talents, teams, and subordinate organizations of all size. Confronted with such kind of complexity overlapping, the human resource department (HRD) frequently get lost and struggle for the elegant order in the practice of enterprise management.

According to the ISO/IEC/IEEE 15288 standard, human resource is an essential part of organizational project-enabling processes [7]. Development is the FIRST job, innovation is the FIRST impetus, and talent is the FIRST resource [8]. So, how to manage the complexity explosion of human resource in an effective and efficient manner, has become a key problem of enterprise management.

Based on the above awareness of human resource management, a model-based approach is proposed to handle the complexity. Even though very few researches have been made in the domain of human resource systems engineering (HRSE), it is still worth a good try. Since the product engineering has taught the staff a lot of MBSE knowledge and modeling skills, it is both necessary and feasible for certain aerospace and aeronautical companies.

2 HRSE Context and Use Cases

HRSE needs to handle tough problems of great complexity. Typically, several following difficulties should be solved.

2.1 Environment Awareness

Nowadays, enterprises live in a world of volatility, uncertainty, complexity and ambiguity (VUCA) [9, 10]. The environment is constantly changing, becoming more unstable each day. As events unfold in totally unexpected ways, it's nearly impossible to predict the root cause or final effect of environmental accident. Circumstances are more complex than ever. Different layers of agents interact each other, making it very hard to get an overtop view of how things are related. This make it quite difficult to choose the single best path under current situation. In business analysis, a PESTEL framework with 6 elements - namely politics, economy, society, technology, environment and legislation - is often used to analyze and monitor the environmental factors [11]. Human resource management issues, such as organizational missions, strategic goals, environmental constraints and macro-level risks, need to be conscientiously defined, adequately communicated and widely accepted. This leads to a good reason for HRSE framework under which each individual and organization may contribute better services or skills in a sense of qualitatively and quantitatively improved results, based on the comprehensive understanding of human resource management system.

2.2 Terminology Confusion

One of the major problems for HRD is the confusion of terminology systems. There are hundreds of terms used to describe the requirements and attributes of workforce. Many of them are measurements of human resources. Misunderstanding of these concepts would cause wrong values and distort the information for decision-making. For example, on-job employee and in-service employee are two completely different concepts while many people know little about their differences. According to the Chinese national formal regulations, for many state-own companies, there may be a difference around 10 percent [12]. If the workforce size misuse a wrong concept, the average efficiency indicators per employee would be calculated by significant error, which could be propagandized to every corner of the enterprise, and eventually confuse the management system. So, it is a critical issue to precisely define the terminologies. Since the terms have associated each other via the definitions, a dictionary of keyword meta-model and a set of models shall be constructed based on a consolidated basis. Many terms are finally complied in a widely accepted dictionary, with which the professional HR team could start their further jobs.

2.3 Pipeline Complexity

The next problem needed to be tackled is the very complicated and interwoven development pipeline system. "One size fit all" has been obsolete nowadays [13], and it is assumed to customized according to various situations. Usually, the workforce has 3 major parts, namely the management team, the science and technology team, and the operating workers. There are more subdivisions for each part, and each subdivision owns a different performance evaluation criteria system with different value grade and weight factors. For instance, the management team includes the leaders and the functional management staff. Their development pipelines vary to each other. Similarly, the science and technology teams are totally different with the operating workers in the salary, career path, honor system, etc. However, under certain circumstances, people can conditionally rotate among the 3 teams. Since the authority wants to maintain a good order of such kind of people smooth flow, the pipeline system need to be designed with strategic intensions and reasonable assumptions [14]. The system is linked to the career fate of every employee, and thus attracts deep concerns. The workforce pipeline system becomes quite complicated, which make MBSE methods and tools quite useful to describe those delicate details.

3 MBSE Approach for Human Resource Management

The above issues related with human resource management are quite complex and very suitable for MBSE methods. A chain of tools is utilized for the human resource systems engineering (HRSE). With the efforts to implement the strategy on developing a high-quality workforce, the enterprise is accelerated to build a major world center of professional talent and innovation. Through the HR management practice in the academy, a set of MBSE models are constructed based on the reality analysis.

3.1 Methodology of HR System

A methodology named by MSFLP – short for mission, strategy, function, logic, and physical -illustrates the overall process of human resource management engineering, as shown in Fig. 1. MSFLP is derived from the MOFLP adopted by MetaGraph. Originally, MOFLP represents mission, operation, functional architecture, logical architecture and physical architecture. The mission starts with defining the commander's intent with linkage to the organizational strategy, guidance, plan, concepts, purposes, or stakeholder inputs. However, in management discipline, strategy is more appropriate than operation, so is used to replace operation. Strategy defines the intended plan for the realization of missions. Functionality stress the design characteristics of system. Logical Design and Physical Design represents the decomposition on different levels. Service on the left side of the architecture and behavior on the right side, mean that HRD need to continuously serve the enterprise and employees through behaviors along all levels.

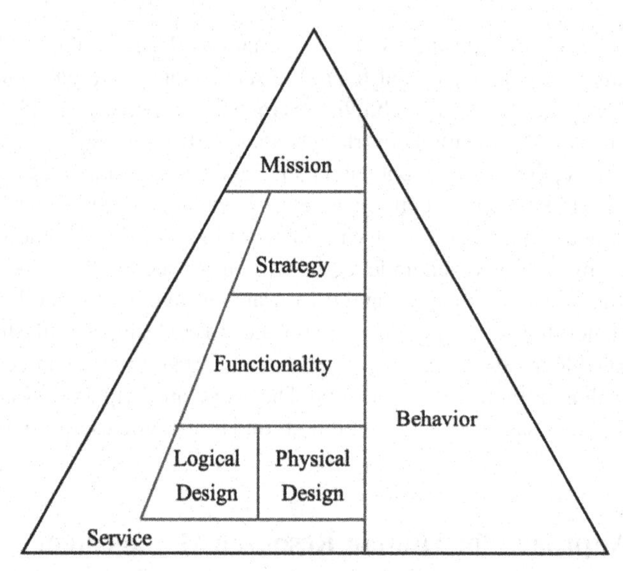

Fig. 1. The MSFLP methodology for HR system engineering

3.2 Mission Definition of HR System

It is the start point to define the mission of human resource management. In the VUCA world, a rigorous use case model can illustrate the mission of HRD. Stakeholders are analyzed in a semantic modeling approach to support the process. PESTEL factors are also integrated into the model. HRD get its strategy aligned with enterprise and national

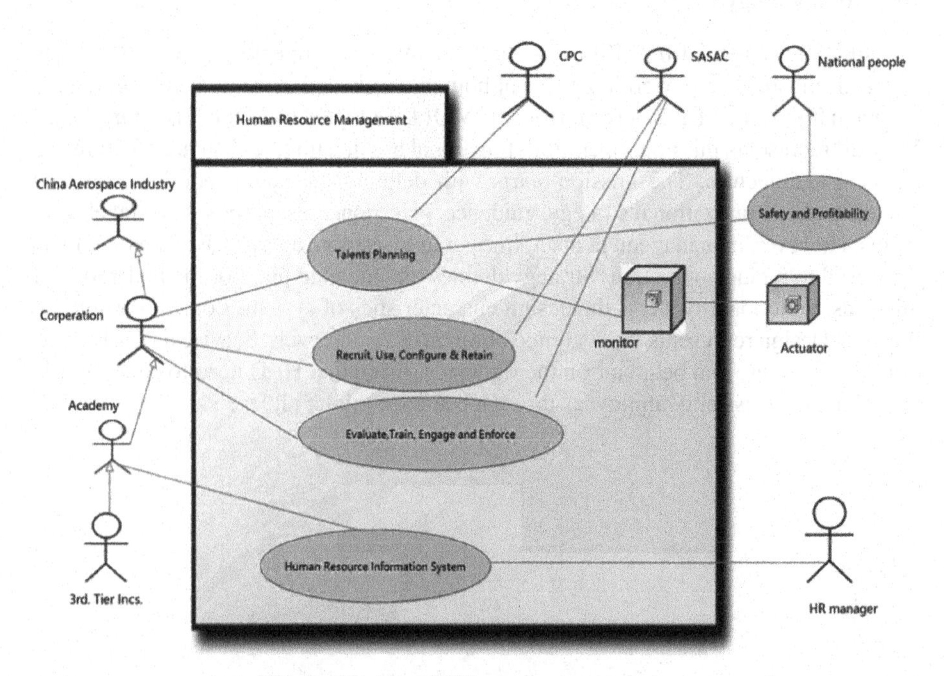

Fig. 2. Use case of HR management mission

interests as shown in Fig. 2, modeled by MetaGraph in KARMA modeling language with SysML specifications. KARMA (Kombination of ARchitecture Model specifi- cAtion) language is a semantic modeling language for a multi-architecture modeling approach [15]. MetaGraph presents metamodels that support the various MBSE formalisms such as SysML across lifecycle stages, providing an ontology based upon graphs, objects, points, properties, roles, and relationships with extensions (GOPPRRE) [16].

During the mission defining process, two major system need to be constructed. One is HR terminology system, which defines the semantic meaning and precise calculation formula. The other is the HR concept model system, which defines the HRD business architecture and hierarchy decompositions.

HR Terminology System

Terms are the basis of the organizational and professional communication as well as cooperation. MetaGraph and KARMA language can define the meta model of HR terminology system and annihilate the misunderstandings [17]. Around 64 crucial keywords, terms, abbreviations, indicators are defined precisely by semantic modeling approach with MetaGraph, as illustrated in as Fig. 3. Merely the text definitions cannot fully illustrate the relationships among the terminologies. Each enrolled professional term owns a definition. If the term is a quantitative indicator/measurement, a calculation formula would also be attached. Moreover, the indicator's detailed statistical scope is clarified. For example, (1) number of employees in service and number of employees on job are clearly defined, (2) the profit ratio of human cost equals to the gross profit divided by the on-job number of employees. Through the format of formal file, the recognitions of professional human management are consolidated. This builds a shared cognition on HR management, and helps a lot in the following enterprise practice.

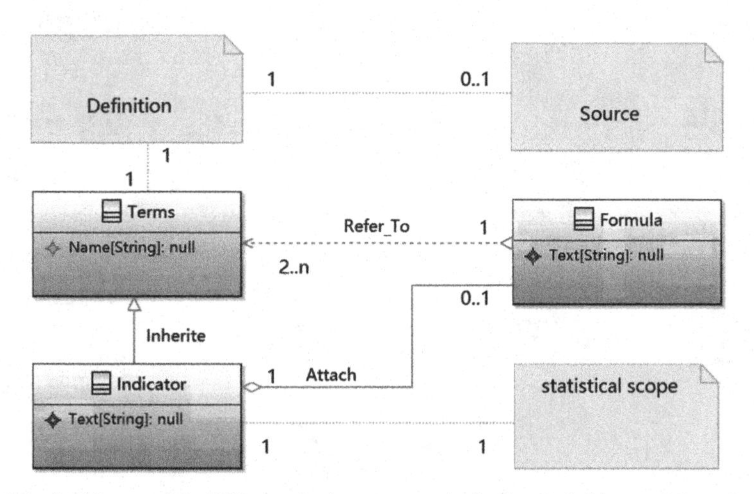

Fig. 3. Metamodel of HR terminology system including definition and formula

HR Concept Modeling

Object Process Methodology (OPM) is good to explain the interactions of different levels' functionalities, especially when the concepts are a little confused or less matured [18]. As Fig. 4 shows, human resource management is constrained by two major inputs: one is the 14th Five-Year general plan which describes the comprehensive strategy of the academy, the other is the 14th Five-Year talent plan of corporation group which talks more on the HR functionality of higher-level organization. The *14th Five-Year Talent Plan* of academy is undertaken by many divisions including planning division, which in turn comprise of two states – *drafting* and *approved*, and a lot of tasks including *Three-System Reforms*, workforce pipeline management, internal talents market, honor system, and human resource information system (HRIS) etc. The oval shape represents the *internal talents marketing*, which is a critical process for both employee in/out management and *head count control* management. Environment objects are represented by boxes with dashed boarders. Different background colors highlight the interested elements of the HR system. So, OPM's rich semantics help to precisely define the concepts, strategies, static structures, notional relationships and related stakeholders. However, the representation of links in MetaGraph still have some inconsistency with ISO/PAS 19450.

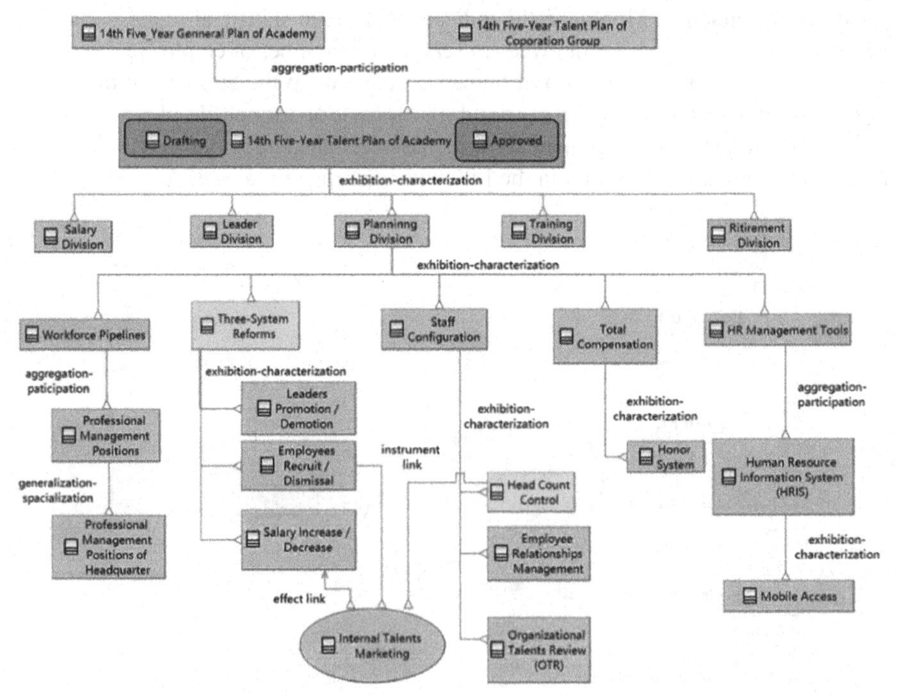

Fig. 4. The OPM model of the 14th Five-Year talent plan (Note: from the view of Planning division of HR department)

3.3 Strategic Analytics Based on System Dynamics Model

System Dynamics modeling tool is useful for explanation of the HR systems' strategy and first principle - such as inflow-stock-outflow model - which is quite self-evident for people. HRD cultivates a talent pool in various areas, such as mangers, engineers, scientists, electrical technician, and so on. The inflow is adjusted to lower the volume desired stock amount, and the outflow is tried to be open wider. At the same time, the enterprise takes a lot of efforts – such as launch new initiatives – to make a full use of stock talents. In short, this model in Fig. 5 makes everyone who involved the human resource management quite aware of the situation and strategy. A talent pool model based on system dynamics helps to enhance the pipeline system improvement, nurturing a large number of top sci-tech talents and teams. Figure 6 depicts a mechanism build by MetaGraph with system dynamics module. Based on system thinking, this model provides deeper insights of factors that impact the head count of workforce. In HRSE, head count is a central element almost linked to all parts. There are 6 entities playing the inflow/outflow roles, respectively 5 from upstream, 1 to downstream. Among the 5 inflow relationships, 3 of them are marked with a negative symbol "–", which means that the improvement of the Three-System Reforms, training hours, and digital transformation can contribute the downsizing of head count. The strategy become very clear toward the head count control objective, contributing a lot for the high-quality development which requires less employees to fulfill more value.

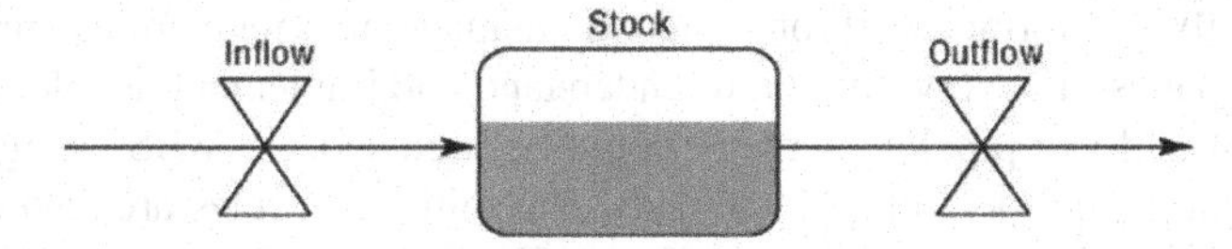

Fig. 5. Inflow-stock-outflow model of system dynamics for HR management

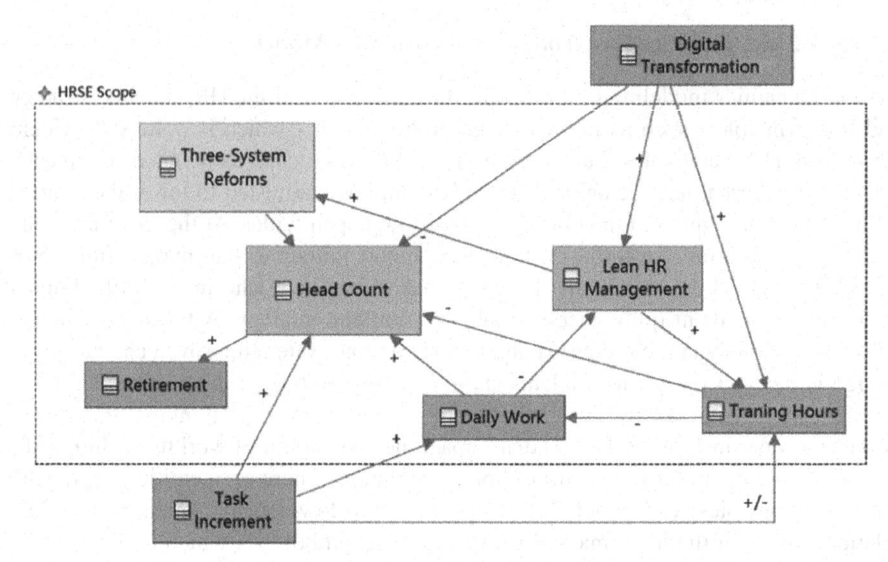

Fig. 6. Dynamics model of head count control system

3.4 Functionality of Workforce Development System

SysML plays a critical role in describe the functional characteristics of well-defined HR system. Usually, the formal workforce pipeline management regulations are written in text of many pages. It is very difficult to understand and implement, let alone optimize the system. Workforce pipeline network could be modeled via MBSE languages like SysML and modeling tools like MetaGraph. The mitigation rules are also quite clear and easy to be understood, as shown in Fig. 7. Furthermore, it is easy to find that the leader pipeline on the left side is much shorter than the professional management pipeline on the right side. For instance, the nominal shortest career path of leader pipeline takes only 12 years from a newbie to chief engineer of the academy, while the nominal shortest career path of professional pipeline takes 43 years to a corresponding level. This can explain why people strive to enter the leader pipeline at no cost, while the professionals confront a shortage in talents. Therefore, the pipeline system should be balanced later based on the analysis. If the workforce development system is carefully optimized, both the organization and the employee would benefit from this model.

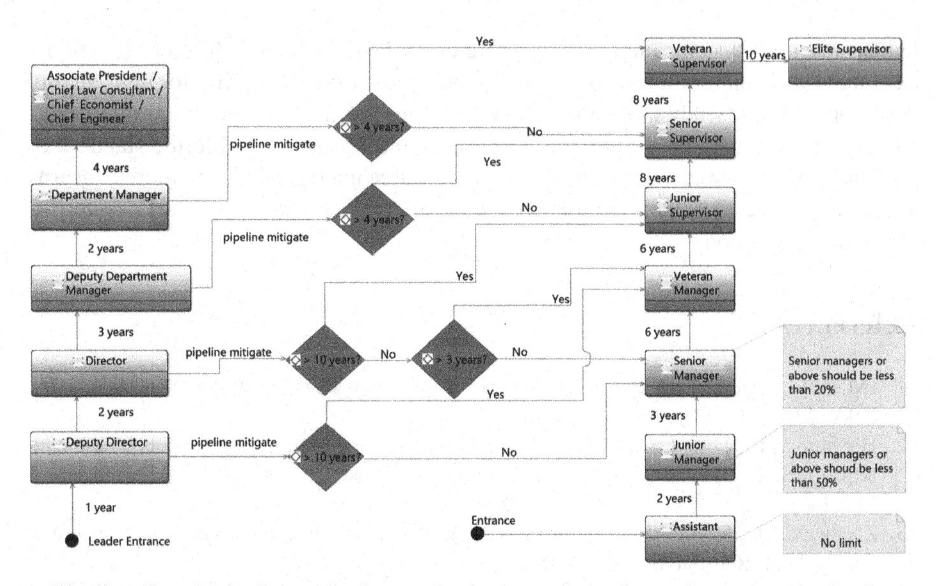

Fig. 7. The as-is pipeline mitigation mechanics between leading cadres and professionals

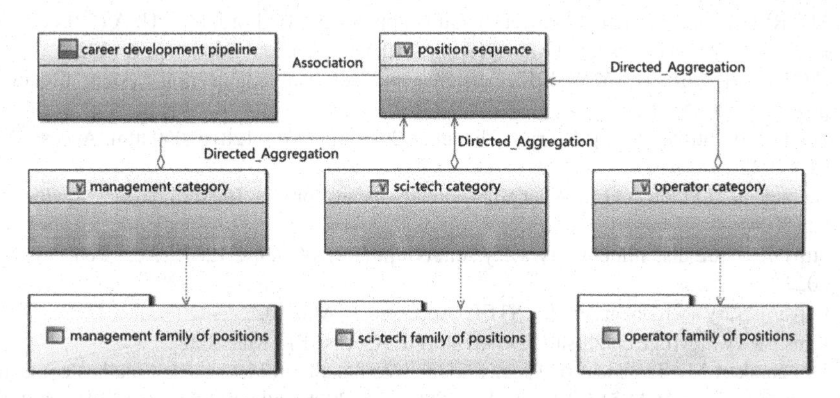

Fig. 8. Workforce position system model

MetaGraph can take a more intensive use recursively when a new complex system – such as workforce position system – need to be defined. As Fig. 8 shows, each pipeline consists a sequence of positions that categorized into 3 types. Each type consists of miscellaneous positions which could be defined or expanded later. Next, logical and physical decompositions can be done in a recursive manner. No matter how complex the position system is, this method can handle the complexity in a feasible approach. The modeling practice also proved the flexible meta-modeling functions of MetaGraph.

4 Conclusion and Future Work

MBSE can be used not only in product engineering, but also in enterprise engineering, such as human resource management. MSFLP architecture is developed to align the human resource systems engineering. Different MBSE tool has a different functional advantage, and finally form up a chain of value to maximize the utility for high-quality development. Case study shows that HR systems engineering conforms the similar applicability of MBSE. In the future, more efforts would be taken to build the standard or referential domain-specific models of human resource management, and more concerns should be shifted to model-based HRSE to accelerate the building of world center for talent and innovation.

References

1. INCOSE: Systems Engineering Handbook, 4th edn, pp. 189–190. Wiley, San Diego, CA, USA (2015)
2. Hirshorn, S.R., Voss, L.D., Bromley, L.K.: NASA systems engineering handbook, 2nd edn. https://ntrs.nasa.gov/api/citations/20170001761/downloads/20170001761.pdf. Accessed 15 Mar 2022
3. Zheng, X., Qu, X.: Development process of Qian Xuesen's thought on systems engineering. Sci. Technol. Rev. **36**(20), 6–9 (2018)
4. Drucker, P.F.: Management Challenges for the 21st Century. China Maine Press, Beijing, China (2020)
5. MITRE: MITRE Systems Engineering Guide, pp. 54–55. Bedford, MA, USA (2014)
6. Dessler, G.: Human Resource Management. 15th edn. Pearson, London, UK (2021)
7. ISO: ISO/IEC/IEEE 15288:2015, Systems and software engineering-System life cycle processes (2015)
8. QS Theory: http://www.qstheory.cn/dukan/qs/2021-12/15/c_1128161060.htm. Accessed 15 Mar 2022
9. Bennett, N., Lemoine, G.J.: What VUCA really means for you. Harvard business review, no. 92 (2014)
10. https://esajournals.onlinelibrary.wiley.com/doi/pdf/10.1002/ehs2.1267. Accessed 15 Mar 2022
11. https://library.sacredheart.edu/PESTEL. Accessed 15 Mar 2022
12. https://www.lawtime.cn/zhishi/a2908518.html. Accessed 15 Mar 2022
13. Stonebraker, M.: Technical perspective. One size fits all: an idea whose time has come and gone. Commun. ACM **51**(12(December 2008)), 76 (2008). https://doi.org/10.1145/1409360.1409379
14. Walcutt, J.J., Epstein, R., Torgler, J.: Designing a defense M&S workforce pipeline to promote national readiness. In: Interservice/Industry Training, Simulation, and Education Conference (I/ITSEC) (2020)
15. Lu, J., Wang, G., Ma, J., Kiritsis, D., Zhang, H., Törngren, M.: General modeling language to support model-based systems engineering formalisms (part 1). In: INCOSE International Symposium, vol. 30, no. 1, pp. 323–338 (2020)
16. Lu, J., Ma, J., Zheng, X., Wang, G., Li, H., Kiritsis, D.: Design ontology supporting model-based systems engineering formalisms. EEE Syst. J. https://doi.org/10.1109/JSYST.2021.3106195
17. Jiangmin, G., Lu, J., Guoxin, W., Shengnan, S., Hang, Z.: General modeling language supporting architecture-driven and code generation of MBSE (Part 2) (2020). https://doi.org/10.1002/j.2334-5837.2020.00797.x.
18. ISO: ISO/PAS 19450 Automation systems and integration–Object-Process Methodology, 15 December 2015. https://www.iso.org/obp/ui/#iso:std:iso:pas:19450:ed-1:v1:en

Complex Systems Modeling and Knowledge Technologies

Knowledge Technology and Systems - Definition and Challenges

Yoshiteru Nakamori(✉)

Japan Advanced Institute of Science and Technology, Nomi, Japan
nakamori@jaist.ac.jp

Abstract. This paper first introduces a value creation spiral model of converting data and information into knowledge and creating valuable ideas, followed by explaining the action guidelines of the four processes of the model. It defines the theories and tools used in these processes as knowledge technologies and considers future challenges in developing knowledge technologies. This paper then considers the knowledge system that promotes interaction between codified and personalized knowledge and creates ideas for solving a specific problem. It discusses the methodology to build a knowledge system that uses mathematical or intelligent knowledge technologies and participatory knowledge technologies in a mutually complementary manner. Finally, this paper introduces a book that is currently under editing. After detailed definitions of knowledge technology and systems, it will describe knowledge technologies that the members of this academic society are developing. We hope that this book will stimulate discussions about the direction and challenges of the Society.

Keywords: Knowledge technology · Mathematical and intelligent technology · Participatory technology · Knowledge systems

1 Knowledge Technology and Challenges

"Technology" here refers "soft technology" that has no physical substance to operate. Human creative activities and decision-making belong to soft technology. People acquire such technologies from experience and validate them through tests and experiments in social life. This section defines knowledge technology that belongs to soft technology and discusses challenges in its development.

1.1 Defining Knowledge Technology

Very generally, knowledge technology can be defined as follows. Knowledge technology builds and operates tools and systems that manage knowledge-intensive activities and deal with the knowledge generated by those activities [1]. But if we focus on the power of knowledge, we can paraphrase this definition. Knowledge is recognition memorized personally or socially, but there is an important reason to use the term knowledge rather than information. Knowledge is the ability to judge. Knowledge can convert data and information into new knowledge. Define knowledge technology, focusing on this ability.

J. Chen et al. (Eds.): KSS 2022, CCIS 1592, pp. 141–148, 2022.
https://doi.org/10.1007/978-981-19-3610-4_10

Knowledge technology is soft technology that underpins the human creative activities of converting data and information into knowledge, creating new ideas based on that knowledge, and validating those ideas. Knowledge technology is a general term for soft technologies contributing to somewhere in the process from data collection to value creation, including information technology, systems technology, and management technology.

This section introduces a value creation spiral model and regards knowledge technology as the technology used in at least one of the processes of the model. Figure 1 shows the OUEI spiral model [2] to create value from data that consists of the following four processes.

1. Observation: converting data or knowledge into information
2. Understanding: converting information into knowledge
3. Emergence: creating ideas from the knowledge
4. Implementation: validating the value of ideas

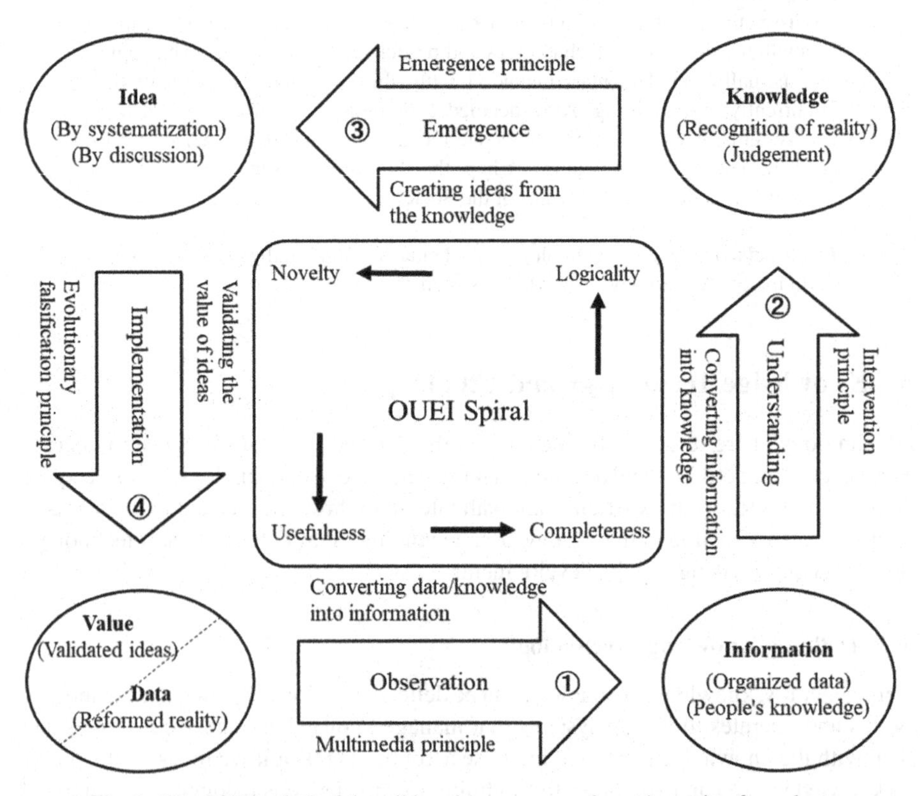

Fig. 1. The OUEI spiral of describing the processes of creating value from data.

First, mobilize all available media to collect the information you need without compromise. Instead of relying on "How-to," pursue "Why-What." Instead of hurrying to produce ideas directly, grasp the essence of the collected information to create knowledge. Produce problem-solving ideas based on the knowledge you have constructed. Keep in mind falsifiability when validating ideas through practice.

Figure 1 presents these cautions as principles to follow in respective processes. They are the multimedia principle, intervention principle, emergence principle, and evolutionary falsification principle. You will also notice that there are four keywords in Fig. 1. They are completeness, logicality, novelty, and usefulness. They represent the goals to be achieved using knowledge technology in the respective processes. See [2] for details.

1.2 Challenges in Developing Knowledge Technologies

This section summarizes [2] to present the challenges in developing knowledge technologies used in the four processes of the OUEI spiral model.

Challenges in Converting Data/Knowledge into Information. When trying to collect information to solve a given problem, you will find two types of information with different origins. One is the data organized for decision-making, and the other is the knowledge of others transmitted by letters, symbols, or voices. Consider the following challenges:

Challenge 1. Develop a measurement optimization framework for collecting just enough quantitative information to explain and solve the problem.

Challenge 2. Develop a hypothesis collection framework that asks "Why" what you are seeing is happening.

Challenges in Converting Information into Knowledge. Knowledge is recognition memorized personally or socially. Notably, knowledge has the power to generate new knowledge from information if it becomes judgments or a judgment system valid objectively. "Knowledge" in knowledge technology has this power. We can divide the knowledge technology that converts information into knowledge into two categories. They are mathematical/intelligent knowledge technology and participatory knowledge technology. Consider the following challenges:

Challenge 3. Advance knowledge technology along the dimension of complexity. Balance the accuracy and understandability of the model.

Challenge 4. Advance knowledge technology along the dimension of human relations. Balance systemic desirability and cultural feasibility.

Challenges in Creating Ideas from the Knowledge. Here, consider two approaches to creating ideas. One is idea generation based on systematized knowledge, and the other is idea generation through people's discussion using creative technologies. The former is the systems thinking approach, and the latter is the knowledge management approach. Consider the following challenges:

Challenge 5. Develop simulation technology for hypothetical reasoning linked with a comprehensive knowledge base and big data.

Challenge 6. Develop knowledge technology that systematically integrates personnel, technology, organization, culture, and management to strengthen the ability to create ideas through knowledge management.

Challenges in Validating the Value of Ideas. Implementation of new products and services is costly and must be internally approved. Once approved, you move on to putting your ideas into practice. In doing so, you consider how to promote and validate your ideas. Consider the following challenges:

Challenge 7. Consider ways to appeal to people's sensibilities, not only to their reason, for justifying and promoting ideas.

Challenge 8. Consider ways to verify ideas through rating your actions that created ideas in addition to quantitative evaluations of ideas.

2 Knowledge Systems and Challenges

Wang [3] proposed the development of a new field of knowledge systems engineering that builds and operates knowledge systems. This section defines the knowledge system and discusses developing the methodology to build knowledge systems.

2.1 Defining Knowledge Systems

Roughly speaking, a knowledge system is a system that creates ideas from data and knowledge. Focusing on the importance of the interaction between explicit and tacit knowledge in knowledge creation, define the knowledge system as follows.

> *A knowledge system is a system that promotes interaction between codified and personalized knowledge and creates ideas for solving a specific problem. Here, codified knowledge shall include data and information, not only knowledge. Personalized knowledge is empirical knowledge or wisdom that is difficult to document.*

A knowledge system includes codified knowledge retainers (i.e., knowledge bases) and personalized knowledge retainers (i.e., the human knowledge resource). Wang and Wu [4] stated, "Due to the abstraction and intangibility of knowledge, people tend to accumulate knowledge from carriers such as books, newspapers, and electronic media. But in essence, people or organizations themselves are careers of knowledge. They compose the fundamental elements of a knowledge system."

The knowledge system does not mean artificial intelligence because it includes carriers of personalized knowledge. However, the elements that behave intelligently with codified knowledge are welcome. Therefore, we use mathematical or intelligent knowledge technologies to systematize knowledge. At the same time, we use participatory knowledge technologies to promote interaction between codified knowledge and personalized knowledge. A knowledge system will achieve its goal through the complementary use of these knowledge technologies. This claim shares the idea with the organizational knowledge creation theory [5] that knowledge emerges through the interaction between explicit and tacit knowledge.

A company is a knowledge system aimed at survival and growth. A large company hierarchically manages many subsystems. Each subsystem needs a talent structure to drive the emergence of ideas to achieve better results. The company must connect upper and lower subsystems properly for smooth communication and control. Each subsystem is a knowledge system with its purpose. A company temporarily forms cross-organizational teams according to their purpose. For example, a new product development team consists of people from R & D, sales, and public relations sections, etc. The team will be a knowledge system that promotes idea creation by systematizing the necessary data, information, and knowledge.

Knowledge technology is meta-knowledge that contributes to creating a part of or all the knowledge system. There is a need for meta-meta-knowledge, or a knowledge system construction methodology, that provides strategies (including usage of individual technologies and their combinations) to develop an entire knowledge system.

2.2 Challenges in Developing Knowledge Systems

To obtain ideas for building knowledge systems, we will learn two systems methodologies: Meta-Synthesis Systems Approach [6, 7] and Critical Systems Thinking [8]. The former promotes a large-scale project by using both quantitative and qualitative methods. The latter suggests appropriate technologies to solve a complex problem. Based on these, this section calls for the development of a systems methodology for building knowledge systems:

Challenge 9. Develop a methodology for building knowledge systems according to the type and scale of the problem.

Building a knowledge system requires using mathematical or intelligent knowledge technologies and participatory knowledge technologies in a mutually complementary manner, as shown in Fig. 2.

Figure 2 assumes that mathematical or intelligent knowledge technologies are responsible for the four processes of the OUEI spiral model: Observation, Understanding, Emergence, and Implementation. The analysts and decision-makers oversee the selection and evaluation of these technologies. They manage the construction of a knowledge system by making full use of participatory knowledge technologies. Knowledge technologies presented in the figure are just examples. In practice, you need to select appropriate knowledge technologies according to the problem. Establishing a methodology for building a knowledge system is a future task.

Consider creating a knowledge system using the knowledge technologies shown in Fig. 2. For instance, imagine a company developing a new product or service. The project team will utilize knowledge technologies as follows.

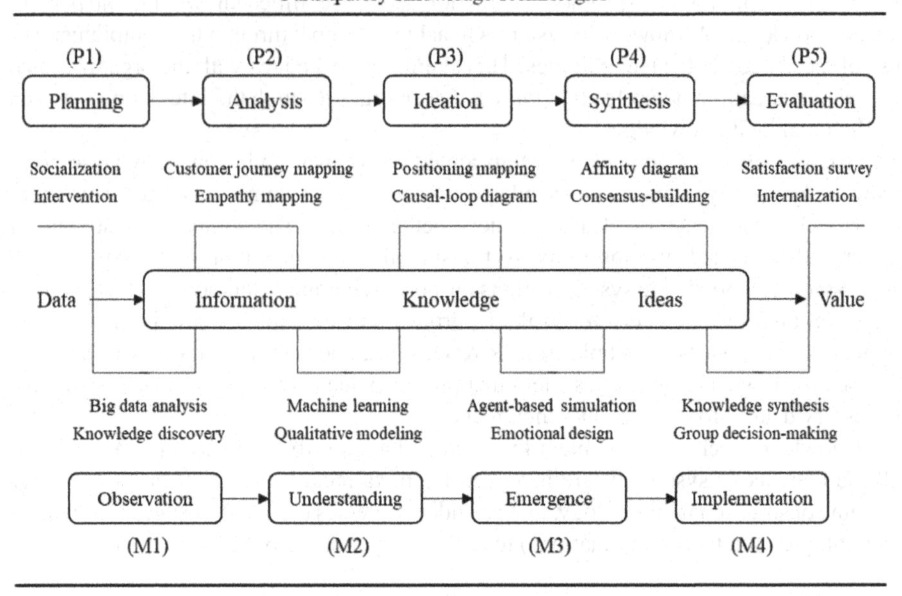

Fig. 2. Knowledge technologies to develop a knowledge system.

(P1) Planning. For the smooth progress of the project, the project leader must carefully listen to the thoughts of management, engineers, sales staff, and, most importantly, consumers. The project team follows the ideas of "socialization" and "intervention" to create a project plan.

(M1) Observation. The information organized from the data usually suggests some meaning. Therefore, some knowledge technologies for converting data into information claim that they extract meaning or even knowledge. The "big data analysis" and "knowledge discovery" are also such knowledge technologies. However, here the project team focuses on the function to convert data into information.

(P2) Analysis. "Customer journey mapping" and "empathy mapping" are technologies that collect evaluation information for products and services. The project team can use the information collected by these technologies in (M2) "qualitative modeling" and (M3) "emotional design." They can also refer to this information in (P4) "consensus-building" and (M4) "group decision-making."

(M2) Understanding. The project team can use various statistical models for demand forecasting. Here, suppose they focus on "machine learning," which automatically processes large amounts of data, and "qualitative modeling," which extracts rules from text data.

(P3) Ideation. It is desirable to perform thought experiments before simulating with the model. Thought experiments can give you ideas for setting rules when performing

(M3) "agent-based simulation." The project team uses "positioning mapping" to locate existing products and services and seek directions for improvement. They can use the "causal loop diagram" to think about the impact of new products and services.

(M3) Emergence. A mathematical model built by past data does not adequately predict demand for products that the company has never offered in the past. Also, it is difficult to predict demand that reflects the emotional changes of consumers. The project team can use "agent-based simulation" and "emotional design."

(P4) Synthesis. The project team needs to group or rank different ideas according to some criteria. They can use the "affinity diagram" for this purpose. They can also use "consensus-building" to put together their opinions if they diverge.

(M4) Implementation. The project team can use "knowledge synthesis" and "group decision-making" technologies if discussions do not converge or if they want to make more rational decisions. They also must discuss promotions to put the idea into practice.

(P5) Evaluation. The project team must assess the project and its results by verifying the accuracy of the model and forecasts and by conducting a "satisfaction survey" or reputation analysis after executing the idea. In addition, they need to accumulate the experience and knowledge gained by the project and the results. The term "internalization" means learning from practice.

3 Detailed Version

The author is editing the following book that details the contents of this paper and explains the mathematical or intelligent knowledge technologies shown in Fig. 2.

Title: Knowledge Technology and Systems

Subtitle: Toward Establishing Knowledge and Systems Science

Editor: Yoshiteru Nakamori

Publisher: Springer

The tentative chapters and contributors are:

1. Defining Knowledge Technology and Systems (Yoshiteru Nakamori)
2. Big Data Analysis in Healthcare (Chonghui Guo, Jingfeng Chen)
3. Knowledge Discovery from Online Review (Jiangning Wu)
4. Machine Learning in Knowledge Society (Wen Zhang)
5. Qualitative Modeling for Problem Structuring (Xijin Tang)
6. Agent-Based Simulation of Environmentally Friendly Product Diffusion (Tieju Ma, Arnulf Gruebler)
7. Emotional Design and Emotional Product Development (Hongbin Yan)
8. Knowledge Synthesis and Promotion (Yoshiteru Nakamori, Siri-on Umarin)
9. Group Decision Making (Jian Chen)

The contributors are researchers of the International Society for Knowledge and Systems Sciences. Individual researchers are developing specified technologies, while the Society covers technologies for the entire process of creating valuable ideas from data. The Society, established in 2003, aims at co-evolving knowledge science and systems science.

Knowledge science can learn the idea of systems science that suggests optimal behavior for the purpose based on systematic and objective information. Systems science, contrarily, can incorporate the methods of knowledge science that utilize the vivid ideas of people.

Review the subtitle of this book. It is "Toward Establishing Knowledge and Systems Science." Notice that "Sciences" has changed to "Science." It tells that we aim to integrate knowledge science and systems science to establish a new field for problem-solving in the age of big data.

References

1. Wang, Z.T.: Knowledge technology. In: Nakamori, Y. (ed.) Knowledge Science: Modeling the Knowledge Creation Process, pp. 11–37. CRC Press, Boca Raton (2011)
2. Nakamori, Y.: Knowledge Technology: Converting Data and Information into Valuable Ideas. Springer, Singapore (2021). https://doi.org/10.1007/978-981-16-3253-2
3. Wang, Z.T.: Knowledge systems engineering: a new discipline for knowledge management and enabling. Int. J. Knowl. Syst. Sci. **1**(1), 9–16 (2004)
4. Wang, Z., Wu, J.: Knowledge systems engineering: a complex system view. In: Nakamori, Y. (ed.) Knowledge Synthesis. Translational Systems Sciences, vol. 5, pp. 107–149. Springer, Tokyo (2016). https://doi.org/10.1007/978-4-431-55218-5_7
5. Nonaka, I., Takeuchi, H.: The Knowledge-Creating Company: How Japanese Companies Create the Dynamics of Innovation. Oxford University Press, New York (1995)
6. Qian, X.S., Yu, J.Y., Dai, R.W.: A new discipline of science—the study of open complex giant system and its methodology. Chin. J. Syst. Eng. Electron. **4**(2), 2–12 (1990)
7. Gu, J.F., Tang, X.J.: Meta-synthetic approach to complex system modeling. Eur. J. Oper. Res. **166**, 597–614 (2005)
8. Jackson, M.C.: Critical systems thinking and practice. Eur. J. Oper. Res. **128**, 233–244 (2001)

Hierarchical Storyline Generation Based on Event-centric Temporal Knowledge Graph

Zhihua Yan(✉) and Xijin Tang

Academy of Mathematics and Systems Science, Chinese Academy of Sciences, Beijing 100190, China
zhyan@amss.ac.cn, xjtang@iss.ac.cn

Abstract. As the main channel for people to obtain information and express their opinions, online media generate a huge amount of unstructured news data every day, which bring great difficulties for people to perceive social events and grasp the development of events. Event-centric knowledge graph has been used to facilitate the reconstruction of news to form structured event information. Most existing studies generate timeline based on event-centric knowledge graphs without considering the complex relations between events. This paper collects news data from Sina platform, constructs event ontology, and builds event-centric knowledge graph with temporal attribute. Afterwords, we propose a novel storyline generation framework with constraints of coherence and coverage. Experiment results show that our method significantly outperforms two baseline approaches.

Keywords: Storyline · Knowledge graph · Event evolution · Community detection

1 Introduction

With the increasing explosion of Internet, online media such as Sina, Yahoo and Weibo generate huge amount of diverse information everyday [1]. These media facilitate the sharing and broadcasting of breaking news to netizen. However, due to sheer volume of information generated by online media, businesses and government cannot gain a summarized view from streaming news easily [2]. Event-centric knowledge graph can facilitate news into structured information [3]. Nevertheless, it is hard for users to gain comprehensive event information from large event-centric knowledge graphs containing millions of events, persons and organizations [4,5]. Timelines are generated from event-centric knowledge graphs to reveal insights of a particular entity [6–8]. While these studies only construct timelines from the perspective of entities without considering complex relations between events. Furthermore, event-centric knowledge graph that contains only simple events is difficult to be used for analyzing societal issues.

J. Chen et al. (Eds.): KSS 2022, CCIS 1592, pp. 149–159, 2022.
https://doi.org/10.1007/978-981-19-3610-4_11

In this study, we propose a novel hierarchical storylines based on event-centric knowledge graphs (HSEKG). In order to gain the evolution of events, this paper extracts time as an attribute of event, and determines the temporal relationship between events. Firstly, we use BERT fine-tuning and graph neural networks (GNN) to extract event and event relations from news text. The attributes of events contain trigger, subject, object, time, and place. The community discovery algorithm is then used to aggregate the events in each time slice into complex events. Finally, storylines are constructed based on the similarities of complex events, taking into account coherence and coverage.

The remainder of this paper is organized as follows. Section 2 describes the event-centric knowledge graph and the generation of storylines. In Sect. 3, we define the problem and propose the framework of this paper. The process of constructing event-centric temporal knowledge graph and generating hierarchical storylines are described in Sect. 4. Section 5 uses a real-world dataset for algorithm evaluation and gives a case study. Finally, we conclude the paper and propose subsequent research directions.

2 Related Works

2.1 Event-centric Knowledge Graph

Knowledge graph is proposed by Google based on the Semantic Web, to enable intelligent search and provide more accurate search results. Knowledge graphs improve the efficiency of knowledge presentation, retrieval and reasoning by representing concepts, common knowledge and their relationships as triples. Knowledge graphs are classified into general knowledge graphs and domain knowledge graphs according to their contents [9,10].

With the development of information retrieval, knowledge graphs are used to describe the evolution of events in the real world, extending entity relationships to logical relationships of events, such as causal and temporal relations. Event-centric knowledge graphs was proposed based on simple event model in 2016 [11]. Compared with traditional knowledge graph, event-centric knowledge graph is generated based on events in real world. Gottschalk and Demidova built a multilingual event-centric knowledge graph that integrated 690 thousand events and 2.3 million temporal relations [12]. Li et al. put forward the concept of event evolutionary graph, and analyzed the logical relations between events, such as temporal relations and casual relations [13]. Zhang et al. developed a large-scale eventuality knowledge graph ASER, which contained activities, states, and entities [14]. Wu et al. proposed an Event-centric Tourism Knowledge Graph based on 18 thousand travel notes from Internet to reveal the temporal and spatial dynamics of tourist's trips [15].

2.2 Story Structure Generation

The research of topic detection and tracking discovers news events, groups them by topics, and track previously news events by attaching related new events into the same cluster [16]. A large number of methods have been proposed to obtain the development of events. Nallapati et al. proposed event threading to capture dependencies among events based on similarity [17]. Yang et al. combined the similarities between events, temporal order, and document distribution to decide the relations between events, and used DAG to represent the structure of event evolution [18]. Mei and Zhai discovered the evolutionary pattern of events by generating word clusters for each time slice and used Kullback-Leibler divergence to discover coherent storylines [19].

These researches only examine pairwise relations between events without considering overall storyline consistency. Metro Map defined coherence and diversity metrics to identify lines of documents by maximize the diversity and coherence of storylines [20]. Xu and Tang proposed risk map to reveal the evolution of societal risk events based on coherence, coverage, and connectivity [21]. Celikkale et al. generated a directed graph based on visual, textual and spatio-temporal features of photo album [22]. Norambuena and Mitra employed landmarks and routes to represent different kinds of events, and designed a narrative map to illustrate the events and storylines [23].

3 Problem Definition and Framework

3.1 Problem Definition

In this paper, we propose a framework for hierarchical storylines, as shown in Fig. 1. The event-centric temporal knowledge graph (EventTKG) provides a structured representation of news events and describes the temporal relationships between events. In order to capture the evolution of events in the event knowledge graph, this paper clusters the nodes to form chronological relationships of complex events. The key concepts are defined as follows:

Definition 1. Event: An event is something that happens at specific time, place and carried out by an individual or organization. Events are associated with real world breaking news. Each news consists of a number of events.

Definition 2. Complex event: Complex events are blocks of events involving the same topic, represented by a group of non-overlapping clusters of events. Complex events link to multiple news with the same topic.

Definition 3. Event-centric temporal knowledge graph: EventTKG provides a global view on events and temporal relations over a period. EventTKG integrates events from news, and enriches it with additional features such as subject, object, time, place and so on.

Definition 4. Storyline: A storyline $\mathcal{S}$ is a sequence of complex events arranged in a chronological order. Complex events are one of the basic constituents of storylines. The directed edge from complex event C_1 to C_2 indicates the temporal relation of braking news.

3.2 Framework of Method

The hierarchical storyline is constructed as shown in Fig. 2, and generally consists of four parts as follows.

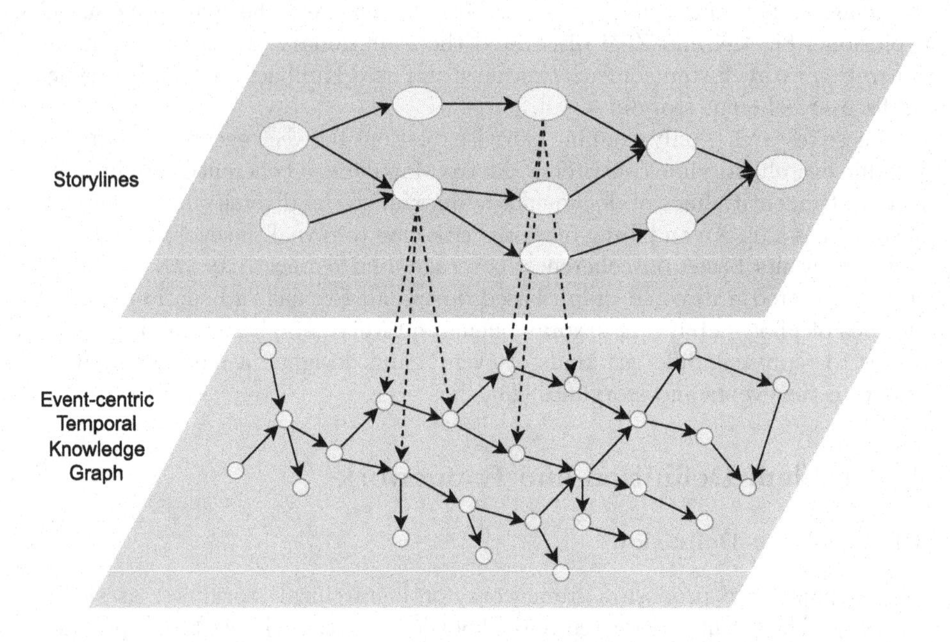

Fig. 1. Hierarchical storylines based on EventTKG.

Firstly, we collect news from Sina platform, which provides real-time coverage and more complete elements of breaking events.

Then data preprocess is employed. As the news contents contain large quantities of noise words, word segmentation is used to reduce the low frequency words. To construct the EventTKG, POS tagging and dependent syntactic analysis are used in event extraction and temporal relation extraction.

Event extraction and event temporal relations extraction are the keys of EventTKG. We use graph neural network with attention mechanism to extract events from sentences of news documents. BERT pre-trained model is employed to extract temporal relations from events in the same news document. Besides, the similarity between event elements is used to solve event coreference problem.

Finally, a novel storylines generation framework HSEKG is proposed, which consists of four components. We leverage community detection algorithm to detect complex events from each time slice. The similarities between complex events is calculated as a coherence metric to generate storylines. The storylines generation are convert into a linear programming problem. Moreover, storylines are visualized to gain more insights from EventTKG.

4 Methods

4.1 Construction of EventTKG

The schema of EventTKG is as shown in Fig. 3. In EventTKG, each event contains four properties: subject, object, time and place. The temporal relation is also included in EventTKG. In Table 1, a brief description of the event attributes is given. We define five attributes to describe each events, including subject, object, place, time and event type.

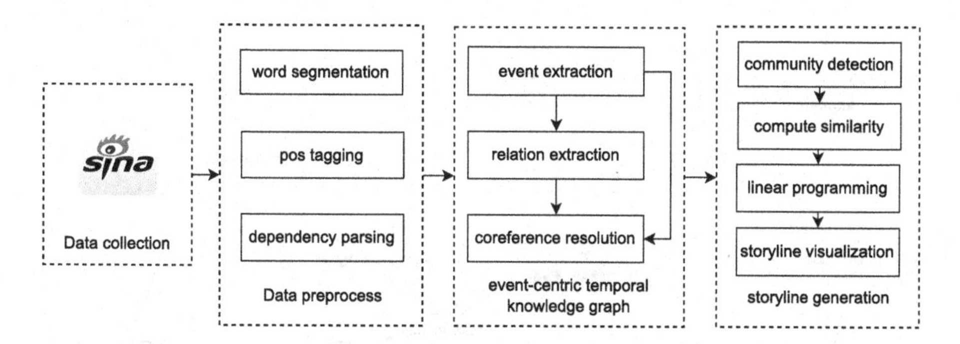

Fig. 2. Framework of HSEKG.

Table 1. The definition of event attributes.

Attribute	Description	Example
Subject	The initiator of the event, including person organization geo-political entity, facility, weapon, etc.	Syria, the foreign ministry, Bush
Object	The patient of the event, including person organization geo-political entity, facility, weapon, etc.	Press conference, the United States
Place	A particular point in space where an event happens	Ukraine
Time	A measured period during which an event exists	April 2020
Event type	A brief summary of event, including military, politics, economy and technology.	Military

In this paper, event extraction method is utilized to extract events from news document. Event triggers, event elements and element roles are identified. An event is usually described by a sentence, and each news document always contains a few events. We divides the news document into sentences and uses BERT fine-tuning technique to identify the entities in each sentence. The types of entities include person, organization, geo-political entity, facility, weapon, etc. Then, dependency syntactic parsing, POS tagging and word embedding are used to construct graph neural network for event detection. We employ word2vec for word embedding, and use Bi-LSTM to gain context information of sentences. The output of Bi-LSTM is feed into graph attention network to reduce the effect of noise words in sentences. Finally, event triggers and event element roles identification are carried out using classifiers.

There are several types of event relation in event-centric knowledge, such as temporal relation, casual relation, hyponymy relation and so on. According to the target of this paper, we only consider the chronological relations of events in the same document. Fine-tuning based on BERT is used to extract event relations. Sentence pairs with candidate events are feed into BERT. Then the vectors of events combining with event position in the sentence are feed into a classifier to determine the relation of the events.

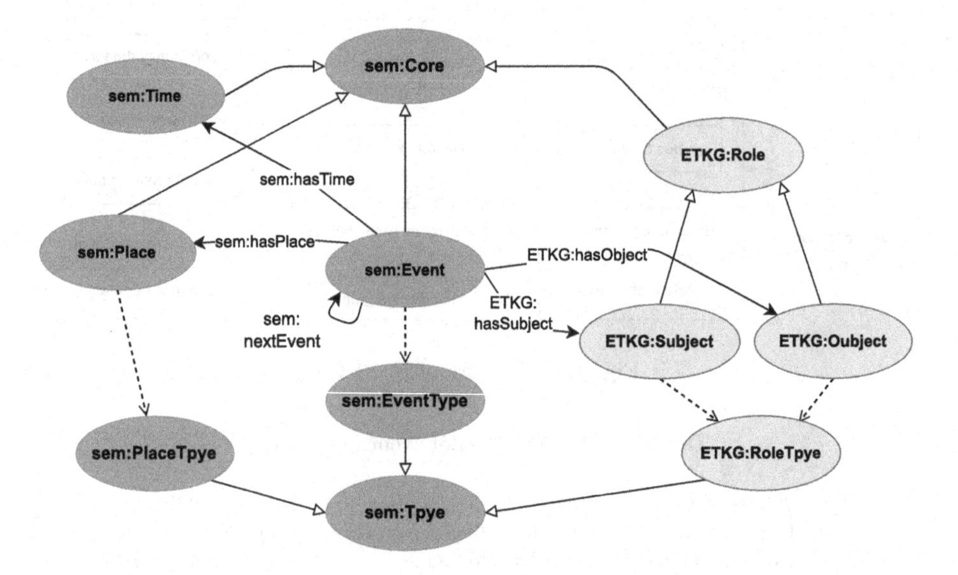

Fig. 3. Schema of EventTKG. Our schema is based on SEM and the classes are colored green. Solid arrows represent regular properties, open-headed arrows represent subclasses, and dashed arrows represent instantiation. Classes and properties introduced in EeventTKG are colored orange. (Color figure online)

Besides, there are a great deal of similar events across the news documents. This paper merges the events with similarities above the threshold. Finally, we store the events and event relations in the neo4j database.

4.2 Generation of Hierarchical Storyline

Given EeventTKG $\mathcal{G} = (V, E)$, V is the set of vertices representing events, and E is the set of edges representing the temporal relations between events. EeventTKG is split into T time slices. In each slice, we use Louvian algorithm to divide the sub-knowledge graph into communities, which denote complex events [20].

In order to generate high quality storylines, this paper introduces the concept of event coherence and coverage. Coherence indicates how well the complex events join together and coverage indicates how well the storylines covers the topics in corpus. As discussed in [3,4], coherence will be high if the complex events in the same storyline. That is, when users follow a storyline along edges joining them, they will get a clear understanding of the underlying theme. Therefore, in order to determine whether it makes sense to connect two complex events with an edge, we need to calculate the similarity of the complex events. Given storylines $S = \{\mathcal{S}_1, \cdots, \mathcal{S}_n\}$, storyline $\mathcal{S}_k = (C_1, \cdots, C_n)$ consist of n complex events, the coherence metric of storyline $\mathcal{S}_k$ is defined as follows:

$$Coherence(\mathcal{S}_k) = \min_{i=1,\cdots,n-1} sim(C_i, C_{i+1}) \tag{1}$$

where $sim(C_i, C_{i+1})$ is the cosine similarity of complex event C_i and C_{i+1}. Hence, the coherence of storylines S is calculated as follows:

$$Coherence(S) = \sum_{\mathcal{S}_k \in S} Coherence(\mathcal{S}_k) \tag{2}$$

The goal of the coverage is to define high coverage storylines, while also being diverse. Coverage should be high if storylines covers as many topics of Eevent-TKG as possible. Prior work relies on simple word frequency-based models and pre-defined keyword lists to compute coverage based on important words given in advance. However, a pre-defined keyword-based approach makes assumptions about how events will increase the risk of selection bias [7]. Thus, to avoid this issue we use LDA model to gain latent topics of the news documents. The topic of complex event is determined by the similarity between complex event and topics. $Cover_s$ indicates the proportion of topics in storyline $\mathcal{S}$. The coverage of the storylines is defined as follows:

$$Cover(S) = 1 - \prod_{\mathcal{S} \in S}(1 - Cover_{\mathcal{S}}) \tag{3}$$

where S is the set of storylines. The higher coverage is, the more diversity of the storylines.

In general, storylines are regarded well if they are maximally coherent for a given coverage requirement. Given storylines $S = \{\mathcal{S}_1, \cdots, \mathcal{S}_n\}$, we compute the optimal storylines by extracting most coherent storylines and selecting a diverse set of topics of complex events by considering the following equations, which are built upon coherence and coverage characteristics:

$$\begin{aligned} S^* &= \underset{\mathcal{S}_i \in S}{\operatorname{argmax}}\, Coherence(\mathcal{S}_i) \\ s.t. &\quad Cover(S) \geq \tau \end{aligned} \tag{4}$$

where τ denotes threshold of coverage score.

5 Evaluation and Case Study

5.1 News Dataset

Currently, there are few open source event datasets. To evaluate HSEKG, we create a Sina News Corpus dataset by collecting news documents from Military Board of Sina. 16,932 news documents are collected throughout 2016. These news cover domestic and international topics, such as politics, military, economy and technology. Then 779 news are labeled manually including events and event relations. Event attributes include trigger, type and subject, object, time and place. Besides, only the temporal relations of events in the same news document are labeled.

On this basis, this paper selects 1370 news documents related to the South China Sea events by keywords, and extracts events and event relations to construct EventTKG. There are 1,395 nodes and 2,381 edges in EventTKG. Besides, we split the dataset into six time slices, and set the threshold of coverage metric 0.7.

5.2 Evaluation

Existing storyline evaluations mainly compute the coherence, coverage, and connectivity of storylines using experts. However, this method has the problem of subjectivity, and the evaluations vary greatly from one person to another. To solve this problem, this paper chooses coherence and coverage metrics discussed in Sect. 4.2 to gain objective evaluation. We choose the datasets about South China Sea events to compare our HSEKG with the following algorithms:

Timeline [25] monitors summary-based variation, volume-based variation and sum-vol variation to produce timelines automatically from tweet streams. A large variation at a particular moment implies a sub-topic change, leading to the addition of a new node on the timeline.

Event Evolution Graphs [18] utilizes the event timestamp, event content similarity, temporal proximity, and document distributional proximity to model the event evolution relations between events in an incident. This method calculates a connection strength for every pair of events and connect the pair if the score exceeds a threshold.

We compare the coherence and coverage metrics of different storyline generation algorithms, as shown in Table 2. As we can see, HSEKG significantly outperforms the other two baseline approaches. Compared with other algorithms, HSEKG uses supervised machine learning to extract events and community detection algorithm to aggregate events into complex events, which are more consistent with the perception of events. As a result, better coherence is found in the storylines generated by HSEKG. In addition, Timeline and Event Evolution Graphs algorithms focus more on the temporal relationships within the storylines and lack the attention to the whole storylines structure.

Table 2. Comparing dierent storyline generation algorithms.

Algorithm	Coherence	Coverage
Timeline	0.641	0.652
Event evolution graphs	0.628	0.628
HSEKG	0.716	0.749

5.3 Case Study

The South China Sea Event is an important event that has received wide attention across world in recent years. Based on the EventTKG, we present a case study to demonstrate how we can leverage the storylines to represent and gain insights as events evolve.

Figure 4 shows four storylines generated by HSEKG with three storylines about diplomatic activities and one about military activity. The nodes represent complex events aggregated by events, and the color of line represents different storyline. The storyline about military activity reveals actions of countries around the South China Sea, such as "Taiping Island", "submarine", "ADIZ", "Self-defense force", "cruise", "aircraft carrier" and so on. The storylines about diplomatic activities have a main storyline, and merge with two other storylines. One storyline is about running runs aground of Philippine fishing boat, and the other is about the Arbitral Tribunal in the South China Sea. The main storyline uncovers the actions of relevant nations, such as China, Philippines, Australia, Asean, Indonesia, Singapore, etc. Through above analysis, we can gain a comprehensive and systematic understanding of the South China Sea event.

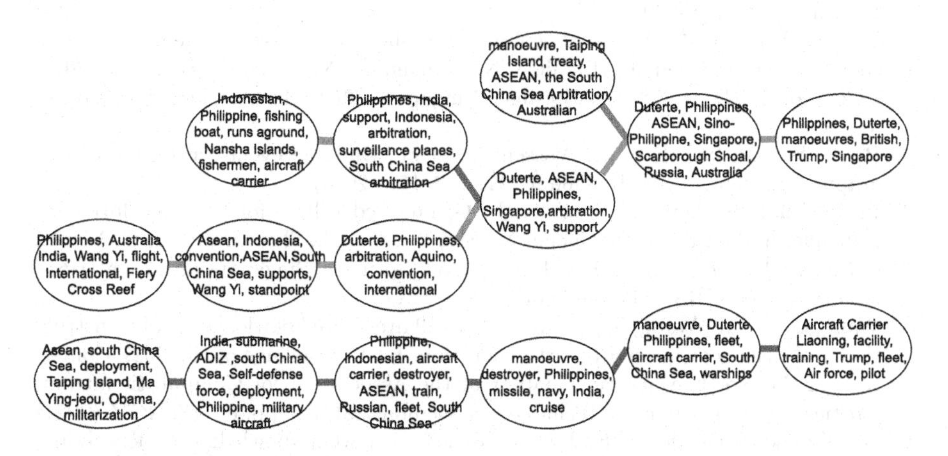

Fig. 4. Case study of hierarchical storylines based on The South China Sea event dataset.

6 Conclusions

This work proposes a novel hierarchical storyline generation framework based on event-centric temporal knowledge graph to reveals the evolution of complex evnents. As we all know, huge number of news generated on media covering politics, economy, society and so on. It is a challenge to uncover the relations of society events and gain the treads of the events. We build an event-centric temporal knowledge graph to get structured representation of events, and generate storylines on aggregated events. The major contributions are summarized as follows:

(1) An event-centric knowledge graph is built with temporal attribute. The temporal relationship is an important feature of event. We propose an event schema with properties of Subject, Object, Time, Place and temporal relation base based on Simple Event Model. This event-centric knowledge graph can reveal complex structures of events extracted from news media.
(2) A novel hierarchical storyline generation framework is proposed. Events in EventTKG are simple events, and unsuitable to analyze societal events. This paper aggregates events into complex events by community detection algorithm. Then storylines are constructed based on the similarities of complex events considering the coherence and coverage.

Acknowledgement. This research is supported by National Natural Science Foundation of China (61473284 & 71971190).

References

1. Dong, T., Liang, C., He, X.: Social media and internet public events. Telemat. Inform. **34**(3), 726–739 (2017)
2. Yan, Z., Tang, X.: Understanding shifts of public opinions on emergencies through social media. In: Chen, J., Huynh, V.N., Nguyen, G.-N., Tang, X. (eds.) KSS 2019. CCIS, vol. 1103, pp. 175–185. Springer, Singapore (2019). https://doi.org/10.1007/978-981-15-1209-4_13
3. Rospocher, M., van Erp, M., Vossen, P., et al.: Building event-centric knowledge graphs from news. J. Web Semant. **37**, 132–151 (2016)
4. Li, Z., Zhao, S., Ding, X., Liu, T.: EEG: knowledge base for event evolutionary principles and patterns. In: Cheng, X., Ma, W., Liu, H., Shen, H., Feng, S., Xie, X. (eds.) SMP 2017. CCIS, vol. 774, pp. 40–52. Springer, Singapore (2017). https://doi.org/10.1007/978-981-10-6805-8_4
5. Wang, Q., Li, M., Wang, X., et al.: COVID-19 literature knowledge graph construction and drug repurposing report generation. In: Proceedings of the 2021 Conference of the North American Chapter of the Association for Computational Linguistics: Human Language Technologies: Demonstrations, pp. 66–77. ACM (2021)
6. Gottschalk, S., Demidova, E.: EventKG-the hub of event knowledge on the web-and biographical timeline generation. Semant. Web **10**(6), 1039–1070 (2019)
7. Gottschalk, S., Demidova, E.: HapPenIng: happen, predict, infer—event series completion in a knowledge graph. In: Ghidini, C., et al. (eds.) ISWC 2019. LNCS, vol. 11778, pp. 200–218. Springer, Cham (2019). https://doi.org/10.1007/978-3-030-30793-6_12

8. Gottschalk, S., Demidova, E.: EventKG+BT: generation of interactive biography timelines from a knowledge graph. In: Harth, A., et al. (eds.) ESWC 2020. LNCS, vol. 12124, pp. 91–97. Springer, Cham (2020). https://doi.org/10.1007/978-3-030-62327-2_16
9. Mahdisoltani, F., Biega, J., Suchanek, F.: YAGO3: a knowledge base from multilingual Wikipedias. In: Proceedings of CIDR 2014 (2014)
10. Lehmann, J., et al.: DBpedia - a large-scale, multilingual knowledge base extracted from Wikipedia. Semant. Web **6**(2), 167–195 (2015)
11. Van Hage, W.R., et al.: Design and use of the simple event model (SEM). J. Web Semant. **9**(2), 128–136 (2011)
12. Gottschalk, S., Demidova, E.: EventKG: a multilingual event-centric temporal knowledge graph. In: Gangemi, A., et al. (eds.) ESWC 2018. LNCS, vol. 10843, pp. 272–287. Springer, Cham (2018). https://doi.org/10.1007/978-3-319-93417-4_18
13. Li, Z., Ding, X., Liu, T.: Constructing narrative event evolutionary graph for script event prediction. In: Proceedings of the 27th International Joint Conference on Artificial Intelligence, pp. 4201–4207 (2018)
14. Zhang, H., et al.: ASER: a large-scale eventuality knowledge graph. In: Proceedings of the Web Conference 2020, pp. 201–211 (2020)
15. Wu, J., Zhu, X., Zhang, C., Hu, Z.: Event-centric tourism knowledge graph—a case study of Hainan. In: Li, G., Shen, H.T., Yuan, Y., Wang, X., Liu, H., Zhao, X. (eds.) KSEM 2020. LNCS (LNAI), vol. 12274, pp. 3–15. Springer, Cham (2020). https://doi.org/10.1007/978-3-030-55130-8_1
16. Allan, J.: Topic Detection and Tracking: Event-Based Information Organization. Springer Science & Business Media, Heidelberg (2012)
17. Nallapati, R., et al.: Event threading within news topics. In: Proceedings of the 13th ACM International Conference on Information and Knowledge Management, pp. 446–453. ACM (2004)
18. Yang, C.C., Shi, X., Wei, C.P.: Discovering event evolution graphs from news corpora. IEEE Trans. Syst. Man Cybern.-Part A Syst. Humans **39**(4), 850–863 (2009)
19. Mei, Q., Zhai, C.: Discovering evolutionary theme patterns from text: an exploration of temporal text mining. In: Proceedings of the 11th ACM SIGKDD International Conference on Knowledge Discovery in Data Mining, pp. 198–207. ACM (2005)
20. Shahaf, D., et al.: Information cartography: creating zoomable, large-scale maps of information. In: Proceedings of the 19th ACM SIGKDD International Conference on Knowledge Discovery and Data Mining, pp. 1097–1105. ACM (2013)
21. Xu, N., Tang, X.: Generating risk maps for evolution analysis of societal risk events. In: Chen, J., Yamada, Y., Ryoke, M., Tang, X. (eds.) KSS 2018. CCIS, vol. 949, pp. 115–128. Springer, Singapore (2018). https://doi.org/10.1007/978-981-13-3149-7_9
22. Celikkale, B., Erdogan, G., Erdem, A., Erdem, E.: Generating visual story graphs with application to photo album summarization. Signal Process. Image Commun. **90**, 116033 (2021)
23. Keith Norambuena, B.F., Mitra, T.: Narrative maps: an algorithmic approach to represent and extract information narratives. In: Proceedings of the ACM on Human-Computer Interaction, pp. 1–33 (2021)
24. Blondel, V.D., et al.: Fast unfolding of communities in large networks. J. Stat. Mech. Theory Exp. **2008**(10), P10008 (2008)
25. Wang, Z., Shou, L., Chen, K., et al.: On summarization and timeline generation for evolutionary tweet streams. IEEE Trans. Knowl. Data Eng. **27**(5), 1301–1315 (2014)

Research on Construction Method of SoS Architecture Knowledge Graph

Yue Zhang, Minghao Li, Xingliang Wang, Yajie Dou, Bingfeng Ge, and Jiang Jiang(✉)

College of Systems Engineering, National University of Defense Technology, Changsha 410000, China

2638930341@qq.com, jiangjiangnudt@163.com

Abstract. System of systems (SoS) architecture data is the foundation of SoS architecture design, modeling, and evaluation. Traditional methods of architecture data collection generally need the involvement of modelers. However, with the increasing complexity of the SoSs, the scale of architecture data also shows explosive growth. There is an urgent need for a method to automatically collect architecture data from multi-source data. This helps engineers collect architecture data more efficiently. In order to collect architecture data quickly and efficiently, this paper proposes a construction method of the architecture knowledge graph. This method takes the architecture description texts as the input data to automatically extract the architecture data. Firstly, based on the architecture meta-model and user requirements, the schema layer of the architecture knowledge graph is constructed. Then, the architecture named entities are recognized based on the Bidirectional Long Shot-Term Memory Neural Network and Conditional Random Fields (BiLSTM-CRF). And the entity relationships are extracted based on the rule matching method. Finally, the architecture knowledge graph is constructed by using the Neo4j database. Taking the disaster rescue SoS as a case, this paper constructs the architecture knowledge graph of the disaster rescue SoS, which verifies the feasibility and practicability of the method. Compared with the traditional manual methods of data collection, the construction method of architecture knowledge graph can automatically collect architecture data from texts more efficiently.

Keywords: SoS architecture · Knowledge graph · Ontology · BiLSTM-CRF · Neo4j

1 Introduction

Proposed by Google in 2012, knowledge graph is a semantic network that shows knowledge nodes and their relationships [1]. Its elements are knowledge triples in the form of "entity - relationship - entity" and "entity - attribute - attribute value". Compared with other knowledge storage methods, knowledge graphs can intuitively show the relationships between entities. The construction process of knowledge graphs includes knowledge extraction, knowledge fusion, and knowledge storage. Among them, knowledge extraction, including named entity recognition, relationship extraction, and attribute

J. Chen et al. (Eds.): KSS 2022, CCIS 1592, pp. 160–172, 2022.
https://doi.org/10.1007/978-981-19-3610-4_12

extraction, is to extract effective information from multi-source data. It is the core work of knowledge graph construction. There are many knowledge extraction methods including rule-based, statistical-based, and neural network-based methods [2]. With the development of artificial intelligence (AI) and data processing technologies, knowledge graph has become the research focus of academia and industry. At present, the knowledge graph is widely used in different fields, such as power domain knowledge graph [3] and medical domain knowledge graph [4].

In the field of system of systems (SoS) architecture research, there is no efficient method of data collection. At present, architecture data is still collected manually. With the increasing complexity of SoSs, the scale of architecture data is growing explosively. On the other hand, most architecture data exist in architecture description texts, and there is little standardized database or models. As a result, traditional methods of architecture data collection expose shortcomings such as low efficiency and waste of resources. There is an urgent need for a method to automatically and efficiently collect architecture data. Therefore, applying AI technologies in the scenarios of SoS architecture data collection, this paper proposes a construction method for the SoS architecture knowledge graph. This method can automatically collect and store SoS architecture data. Taking the architecture description texts as input data, knowledge extraction technologies are used to collect standard architecture data. And the SoS architecture emphasizes the relationships between SoS components. Storing architecture data in graph databases can intuitively show the relationships between the SoS components. The SoS architecture knowledge graph can provide data support to subsequent tasks such as architecture design and architecture modeling.

2 Construction Framework of Architecture Knowledge Graph

The logical level of the knowledge graph includes schema layer and data layer [5]. The schema layer defines the core concepts such as entity types, relationship types, and attribute types through constructing architecture knowledge ontology. Based on architecture knowledge ontology, architecture data are extracted from texts by using the Bidirectional Long Shot-Term Memory Neural Network and Conditional Random Fields (BiLSTM-CRF) deep learning model [6, 7] and the rule-based relationship extraction method. After collecting data, the architecture knowledge graph is constructed based on the Neo4j graph database [8] to store the architecture data. There are four steps of the framework: data acquisition and processing, schema layer construction, data layer construction, and knowledge storage, as shown in Fig. 1.

(1) Data acquisition and processing. This study uses crawlers to crawl architecture description texts from various websites. These texts mainly include architecture reports and architecture modeling literature. Data cleaning, segmentation, and BIO labeling is carried out for these text data to facilitate subsequent use.
(2) Schema layer construction. Because the architecture data are required to be standard and reusable, the "top-down" method [9] is used to construct the schema layer of architecture knowledge graph. The schema layer ensures the standardization of architecture data by constructing architecture knowledge ontology. Based on the

US Department of Defense Architecture Framework (DoDAF) meta-model [10], the core concepts contained in the architecture knowledge ontology are clarified. Meanwhile, according to the requirements of users, the concepts are perfected to form a complete architecture knowledge ontology.

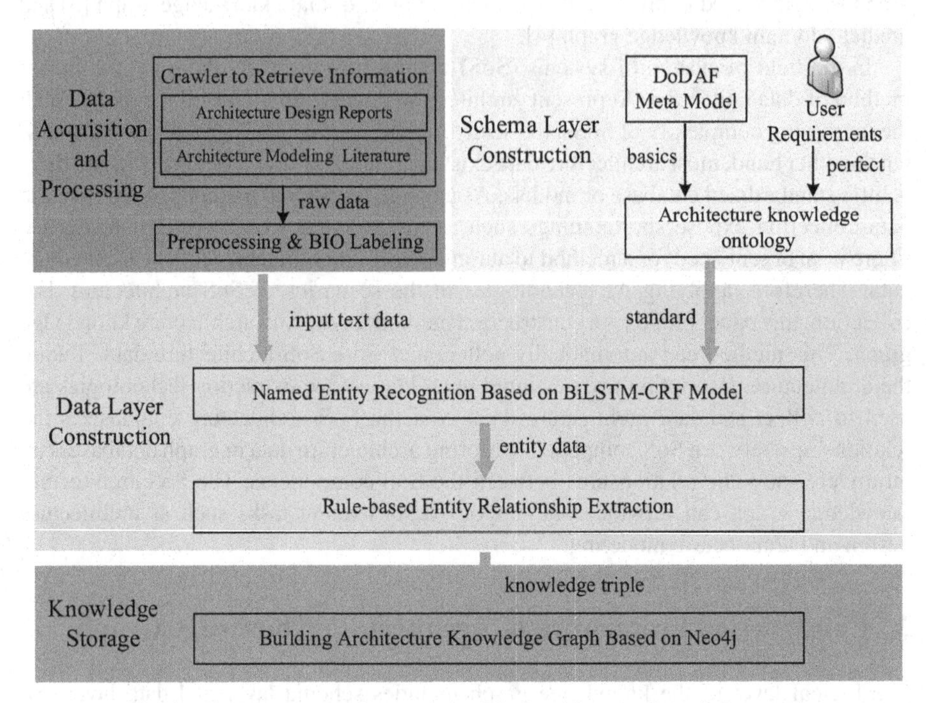

Fig. 1. The construction framework of SoS architecture knowledge graph

(3) Data layer construction. Following the specification of the schema layer, relevant entities are recognized from the architecture description texts based on BiLSTM-CRF deep learning model. After obtaining the entity information, the relationships between entities are collected based on the rule-based relationship extraction method. Attribute extraction can be regarded as a special type of relationship extraction [11].
(4) Knowledge storage. Based on the Neo4j graph database, the architecture knowledge graph is constructed to store the architecture data. The architecture knowledge graph can intuitively show the relationships between entities.

3 Construction Process of Architecture Knowledge Graph

3.1 Data Acquisition and Processing

The input data needed in the construction of the architecture knowledge graph are architecture description texts, which belong to unstructured data. Crawlers are used to crawl the reports and documents related to the SoS architecture from websites to collect the input data. After data cleaning, 7961 usable sentences are collected. Then, these sentences are segmented according to punctuation.

In addition, the processed data need to be labeled to support the training of the BiLSTM-CRF model. In this paper, the BIO labeling method [13] is adopted. BIO labeling method is to label each element as "B-X", "I-X" or "O". "B-X" means that the fragment where this element is located belongs to type X and this element is at the beginning of this fragment. "I-X" means that the fragment where this element is located belongs to type X and this element is in the middle of this fragment. "O" means that the element does not belong to any type. Based on the architecture ontology, the label types are determined, such as "B-Person", "I-Person" and "B-Organization". 7961 text data are labeled manually. Part of the labeling results is shown in Fig. 2.

The	O
general	B-ORG
command	I-ORG
station	I-ORG
commands	O
MAV	B-SYS
flight	O
MAV	B-SYS
and	O
UAV	B-SYS
need	O
to	O
perform	O
cooperative	B-ACT
detection	I-ACT
tasks	O

Fig. 2. Part of the BIO labeling results

3.2 Schema Layer Construction

The architecture knowledge graph is a kind of domain knowledge graph. So, the construction of the architecture knowledge graph must clarify the knowledge types and data format. Constructing architecture knowledge ontology can ensure that the architecture data collected by knowledge extraction is standardized. In the field of architecture research, meta-model theory has matured to guide the collection, organization, and storage of architecture data [12]. This is similar to the normative effect of the schema layer of the knowledge graph. Therefore, constructing architecture knowledge ontology based on the architecture meta-model can greatly improve the efficiency of ontology construction and ensure the standardization and completeness of ontology.

However, the architecture knowledge ontology based on the architecture meta-model is still abstract. The knowledge ontology needs to be perfected according to the specific requirements of users. For example, according to the architecture meta-model, data types such as "capability", "activity" and "system" in the architecture knowledge ontology can be determined. However, in the target system, "system" may include "hardware system" and "software system". And these two subtypes of "system" have specific attributes respectively. Therefore, we should perfect the data types and entity attributes in the ontology according to the requirements of users. Part of the core concepts of architecture ontology is shown in Table 1.

Table 1. Part of the core concepts of architecture ontology.

Core concept	Description
Concept type	
Capability	The ability to employ various means to achieve desired results
Activity	The act of converting input resources into output resources or changing the state of resources
System	A collection of elements that have functional, physical, or behavioral interactions
Person	Personnel type defined according to the tasks undertaken by the personnel
Material	Equipment and supplies. It is not distinguished by its purpose
...	...
Relationship type	
Capability - include - capability	The whole-part relationship between capabilities. A capability may have sub capabilities
Capability - support - activity	The overlapping relationship between capabilities and activities. Activities are supported by the determined capabilities
System - perform - activity	The overlapping relationship between systems and activities. Activities are performed by the determined systems
Person - compose - system	The whole-part relationship between persons and systems. A system may include persons
...	...

3.3 Data Layer Construction

Architecture Named Entity Recognition Based on BiLSTM-CRF. It mainly recognizes named entities such as person, place, organization, and time from text data. Named entity recognition based on neural network is the most popular method at present. The named entity recognition workflow based on the BiLSTM-CRF model is shown in Fig. 3. Using word embeddings to represent words can learn more detailed feature information. This paper selects the model combining Bi-LSTM and CRF to recognize architecture entities. Being able to capture context information, Bi-LSTM is suitable for modeling time series data such as text data. However, if only Bi-LSTM is adopted, illegal outputs may occur in the prediction results, such as "B-Person I-Organization" and "O I-Person". CRF can learn constraints autonomously through the training process to effectively reduce illegal inputs. Therefore, the BiLSTM-CRF model can better recognize the architecture named entities. The model architecture of BiLSTM-CRF is shown in Fig. 4.

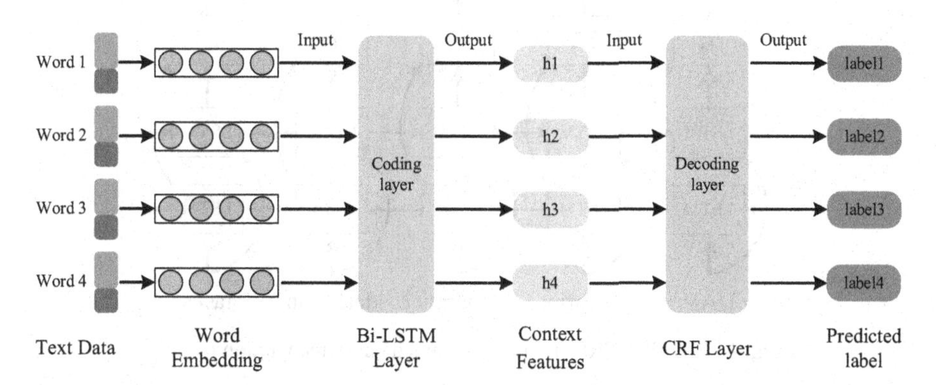

Fig. 3. Architecture named entity recognition workflow based on BiLSTM-CRF

In the process of data acquisition and processing, 7961 text data have been manually labeled with BIO labels. Based on the labeled data, the BiLSTM-CRF model is trained. During training, hide_dim is set to 300, and epoch is set to 12. In addition, the F value is used as the evaluation index of the named entity recognition effect [14]. Its calculation formula is:

$$F = \frac{2 \times P \times R}{P + R} \tag{1}$$

$$P = \frac{Number\ of\ correct\ entities\ recognized}{Number\ of\ entities\ recognized} \tag{2}$$

$$R = \frac{Number\ of\ correct\ entities\ recognized}{Number\ of\ total\ entities} \tag{3}$$

Because the training may have overfitting phenomena, the model with the highest F value will be saved. The F value of the saved model is 79%. Although the trained model has a good named entity recognition effect, the extraction effect can be better. We will continue to train the BiLSTM-CRF model and conduct more experiments to improve the effect of architecture named entity recognition.

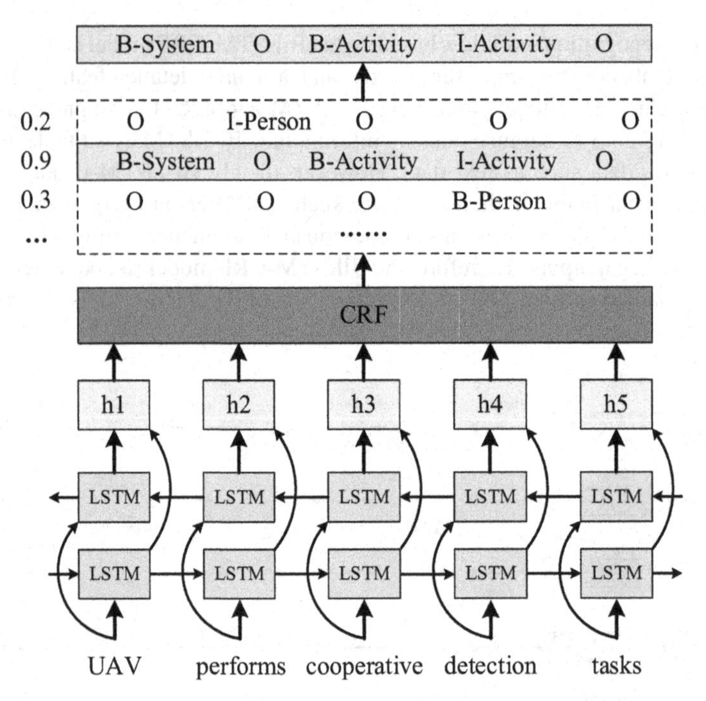

Fig. 4. BiLSTM-CRF architecture named entity recognition model

Relationship Extraction Based on Rule Matching. Relationship extraction is to extract the entity relationships on the basis of entity information. In this paper, the rule-based method is used to extract architecture relationship data. This method needs to make extraction rules based on the semantic features of relationship types. And then match the texts with these extraction rules. Rule-based relation extraction shows a good effect in some professional fields [15, 16]. Firstly, based on the SoS architecture knowledge ontology, the semantic features of relationship types are analyzed to determine the relationship words of various relationship types. And the architecture relationship extraction rules are made. For example, the "include" relationship type between "systems" may be expressed in the form of "system A includes system B", "system B constitutes system A" and "system B is a part of system A" and so on. Therefore, relationship extraction rules for the "include" relationship between "system" are made, as shown in Fig. 5.

[System] include [System];
[System] compose [System];
[System] is part of [System];
......

Fig. 5. Extraction rules of "include" type relationship

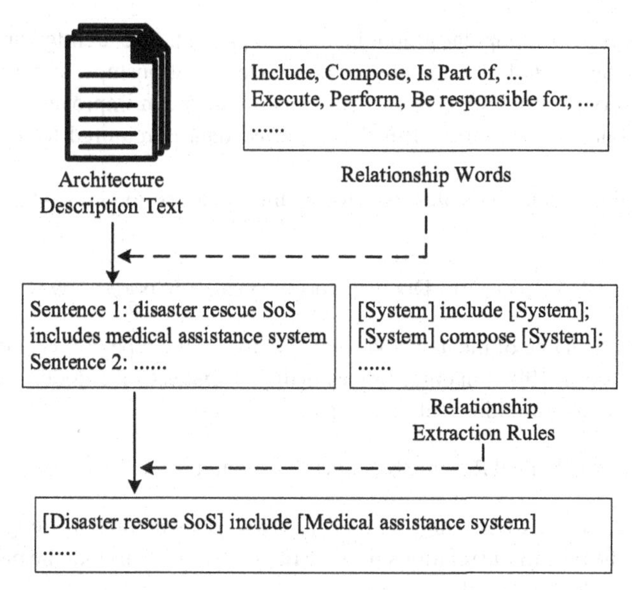

Fig. 6. The process of architecture relationship extraction

After manually making all the architecture relationship extraction rules, the text data are matched with the extraction rules. The architecture relationship extraction process is shown in Fig. 6. Firstly, the sentences containing relationship data are recognized. Then, the architecture entities before and after the relationship word are recognized. Finally, the relationship data conforming to the rules are extracted. Part of the relationship extraction result is shown in Table 2.

Table 2. Part of relationship extraction results

Head entity	Tail entity/Attribute value	Relationship/Attribute
Medical rescue capability	Victim treatment	Capability-support-activity
Command department	Decision making capability	System-have-capability
Navigation satellite	Target location	System-perform-activity
Transport helicopter	3 Tons	Maximum payload
...	...	...

3.4 Knowledge Storage

Storing data based on graph databases can highlight the relationships between entities. Therefore, after collecting the data triples, the architecture knowledge graph is constructed to store the architecture data. In this paper, the Neo4j graph database is used to construct the architecture knowledge graph. Neo4j is a high-performance non-relational database. It is the most popular graph database at present. The architecture knowledge graph can be constructed by manually adding nodes, relationships, and attributes data. In addition, Neo4j supports batch import of data, which can improve the construction efficiency of knowledge graphs. But the imported data is required to have a specific format.

An example of function statements for adding nodes to the knowledge graph is as follows:

$$\text{CREATE(David : Doctor\{Working years : 6, born : 1995\})}$$

"David" is the name of the new node and "doctor" is the type of this node. "Working years: 6" and "born: 1995" in curly braces is the attributes of the node. An example of function statements for adding relationships is as follows:

$$\text{(Medical Rescure System)} - [\text{: Include}] \rightarrow \text{(David)}$$

"Medical Rescue System" and "David" are two nodes in the knowledge graph. "Include" means the type of the relationship. The relationship means that the medical rescue system includes doctor David.

The architecture knowledge graph can support a variety of specific applications. On the one hand, the architecture knowledge graph supports the query of architecture knowledge and the mining of hidden knowledge. On the other hand, the architecture knowledge graph construction method provides a way to automatically collect architecture data, which can directly provide data support for the subsequent work such as architecture design and architecture modeling.

4 Case Application

SoS is a whole of several distributed systems to achieve a certain goal [17]. Its component systems have independence and loose coupling. And the component systems and internal relationships of the SoS are flexible and changeable. The same system may play different roles in different SoSs. Therefore, the commonality of the architecture knowledge graph is not strong. Constructing architecture knowledge graphs for different SoSs can achieve better results. Taking the disaster rescue SoS as a case, this paper constructs the architecture knowledge graph of the disaster rescue SoS. The raw data come from paper [18]. Because the raw data is in the form of text, it is inconvenient to display. 216 text data are obtained after data processing.

According to the construction method of the architecture knowledge graph proposed in this paper, firstly, it is necessary to construct the knowledge ontology of disaster rescue SoS. Based on the DoDAF meta-model, the entity types and relationship types in knowledge ontology are clarified. For example, the knowledge ontology of the disaster rescue SoS should include entity types such as "activity", "system" and "capability", as well as relationship types such as "capability - support - activity" and "system - perform - activity". Then, the knowledge ontology is perfected according to the features of the disaster rescue SoS. For example, the "system" entity type in the disaster rescue SoS includes subtypes such as "hardware system" and "software system". The "person" entity type includes subtypes such as "commander", "medical worker" and "search and rescue worker". And different entity types have different attributes, which also need to be added. The schema layer of the architecture knowledge graph of the disaster rescue SoS is shown in Fig. 7 (Table 3).

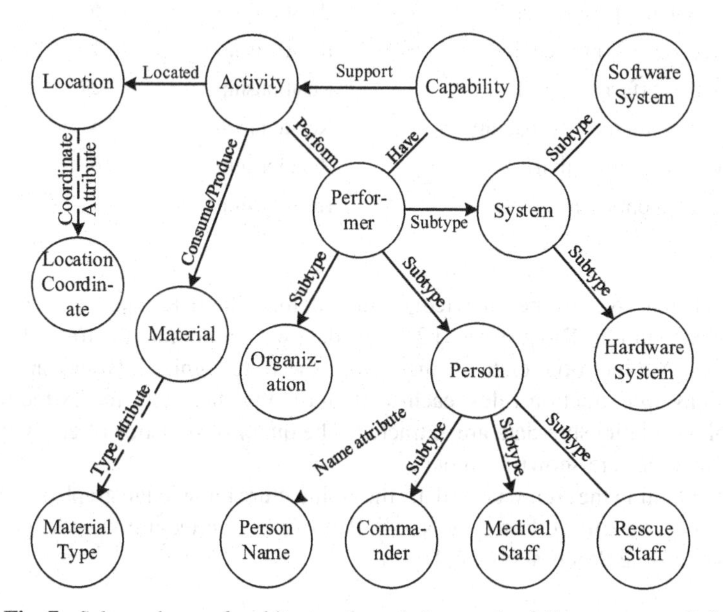

Fig. 7. Schema layer of architecture knowledge graph of disaster rescue SoS

Table 3. Statistics of entities and relationships of disaster rescue system

Name	Type	Quantity
Capability	Entity	9
Activity	Entity	17
Hardware system	Entity	2
Software system	Entity	1
Organization	Entity	6
Rescue staff	Entity	7
Commander	Entity	3
Medical staff	Entity	5
Location	Entity	2
Material	Entity	11
Activity - located - location	Relationship	3

(continued)

Table 3. (*continued*)

Name	Type	Quantity
Activity - consume/produce - material	Relationship	17
Hardware system - perform - activity	Relationship	6
Medical staff - perform - activity	Relationship	19
Rescue staff - perform - activity	Relationship	27
Hardware system - have - capability	Relationship	4
Capability – support - activity	Relationship	6
Medical staff - competent	Relationship	2

The schema layer of the knowledge graph defines the data types to be included in the knowledge graph. The processed 216 text data were input into the trained BiLSTM-CRF model. A total of 63 related named entities were recognized. Based on entity data and relationship extraction rules, each text data is matched with the extraction rules. A total of 84 relationship data are extracted. The quantity statistics of entity types and relationship types are shown in Table 2.

After obtaining the architecture data, the architecture knowledge graph of the disaster rescue SoS is constructed based on Neo4j. Part of the architecture knowledge graph of the disaster rescue SoS is shown in Fig. 8.

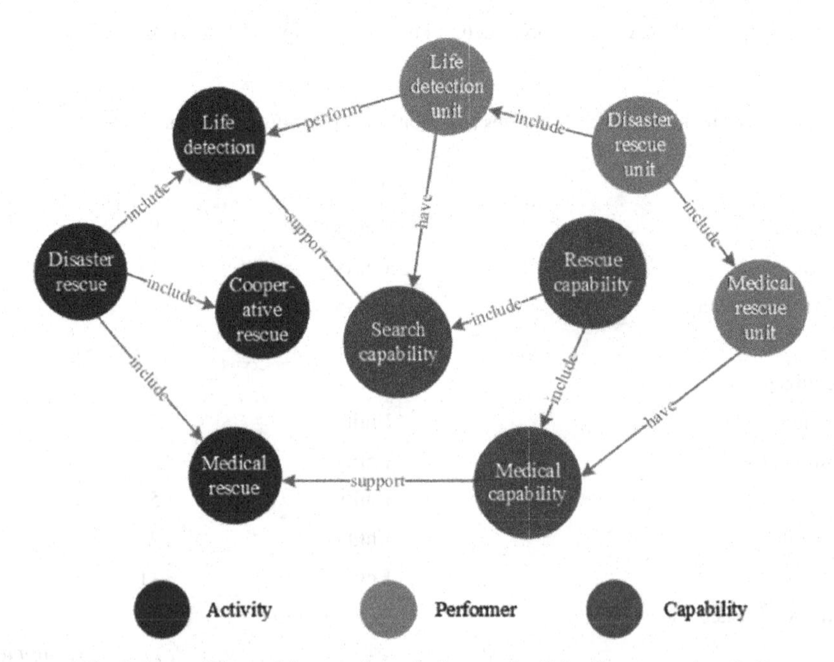

Fig. 8. Part of the architecture knowledge graph of the disaster rescue SoS

5 Conclusion

This paper proposes a construction method for the architecture knowledge graph. Based on ontology construction and knowledge extraction technologies, the method achieves the automatic collection of architecture data from texts. The method has higher efficiency than the traditional manual method. And the architecture data are stored in the knowledge graph, which can provide data support for the subsequent work such as architecture design and architecture modeling. The method promotes the intelligent development of architecture data collection.

Meanwhile, there are still some shortcomings in this paper, which need further research. On the one hand, in the constructing process of the schema layer, architecture knowledge ontology with certain universality has been formed. But the ontology is not complete enough. It needs to be updated constantly to meet the construction requirements of different SoSs architecture knowledge graphs. On the other hand, this paper recognizes the architecture named entities based on the BiLSTM-CRF model. Due to the imperfect quality and scale of training data, the F value is 79%. The BiLSTM-CRF model needs further training to achieve a better extraction effect.

References

1. Bilal, A.: Domain-specific knowledge graphs: a survey. J. Netw. Comput. Appl. **185**, 103076 (2021)
2. Wang, H., Qin, K., Zakari, R., Lu, G., Yin, J.: Deep neural network-based relation extraction: an overview. Neural Comput. Appl. **34**, 4781–4801 (2022)
3. Pu, T., Tan, Y., Peng, G., Xu, H., Zhang, Z.: Construction and application of knowledge graph in the electric power field. J. Power Syst. Technol. **45**(6), 2080–2091 (2021)
4. Yuan, K., Deng, Y., Chen, D., Zhang, B., Lei, K.: Construction techniques and research development of medical knowledge graph. J. Appl. Res. Comput. **35**(7), 1929–1936 (2018)
5. Tian, L., Zhang, J.C., Zhang, J.H., Zhou, W., Zhou, X.: Knowledge graph survey: representation, construction, reasoning and knowledge hypergraph theory. J. Comput. Appl. **41**(8), 2161–2186 (2021)
6. Yan, R., Jiang, X., Dang, D.: Named entity recognition by using XLNet-BiLSTM-CRF. Neural Process. Lett. **53**(5), 3339–3356 (2021). https://doi.org/10.1007/s11063-021-10547-1
7. Krishna, C., Qian, C.: Deep ran: attention-based Bi-LSTM and CRF for ransomware early detection and classification. J. Inf. Syst. Front. **23**(2), 299–315 (2021)
8. Naga, D.: Book recommendation system using Neo4j graph database. Int. J. Anal. Exp. Modal Anal. **12**(6), 498–504 (2020)
9. Huang, H., Yu, J., Liao, X., Xi, Y.: Review on knowledge graphs. Comput. Syst. Appl. **28**(6), 1–12 (2019)
10. Holly, A.: Incorporating the NATO human view in the DoDAF 2.0 Meta model. Syst. Eng. **15**(1), 108–117 (2012)
11. Liu, Q., Li, Y., Duan, H., Liu, Y., Qin, Z.: Knowledge graph construction techniques. J. Comput. Res. Dev. **53**(3), 582–600 (2016)
12. Wan, H., Shu, Z., Huang, L., Luo, X.: Theory of meta model and its application in the development and design of enterprise architecture. Syst. Eng. Theory Pract. **32**(4), 847–853 (2012)
13. Tian, Z., Li, X.: Research on Chinese event detection method based on BERT-CRF model. Comput. Eng. Appl. **57**(11), 135–139 (2021)

14. Duan, H., et al.: Research on water conservancy comprehensive knowledge graph construction. J. Hydraul. Eng. **52**(8), 948–958 (2021)
15. Asma, B., Pierre, Z.: Automatic extraction of semantic relations between medical entities: a rule based approach. J. Biomed. Semant. **2**(5), 4 (2011)
16. Ravikumar, K.E., Majid, R., Liu, H.: BELMiner: adapting a rule-based relation extraction system to extract biological expression language statements from bio-medical literature evidence sentences. Database **2017**(1), 1–12 (2017)
17. Eung-jo, D.: Race to the algorithmic warfare: an analysis of the China's system confrontation and the USA's systems warfare. Strateg. Stud. **28**(3), 217–264 (2021)
18. Li, Y., Li, X., Liu, D.: Architecture models of major disaster rescue system based on DoDAF. Command Control Simul. **38**(6), 16–21 (2016)

Intelligent Modeling Framework for System of Systems Architecture Based on Knowledge Graph

Yue Zhang, Minghao Li, Xingliang Wang, Yajie Dou, Bingfeng Ge, and Jiang Jiang(✉)

College of Systems Engineering, National University of Defense Technology, Changsha 410000, China
2638930341@qq.com, jiangjiangnudt@163.com

Abstract. Enterprise architecture framework such as DoDAF has become an effective method in recent years to describe a system of systems structure and guide its revolution, especially in the military field. However, the lack of professional modelers and domain knowledge has become a stumbling block to the implementation of this management technique. To lower the threshold level for modelers, this paper proposes an intelligent modeling framework that combines ontology-based text recognition and knowledge graph technology to automatically generate architecture models from texts. Compared with traditional rule-based methods, this study has stronger generality and can be applied to more flexible texts. At the same time, it also promotes the application of intelligent research in the field of system of systems management.

Keywords: SoS architecture · Knowledge graph · Intelligent modeling · Framework

1 Introduction

In the military field, system of systems (SoS) confrontation has become the basic form of modern warfare [1]. Combat SoS is an integrated whole of many combat elements. It puts more emphasis on the connections between combat elements. In the enterprise management field, enterprise architecture plays a key role in resource allocation, development plan-making, and others. Most SoSs have complex architecture, which needs to be described by constructing architecture models from different viewpoints. At present, a variety of architecture frameworks have been proposed. Following the specification of the architecture framework, the architecture models are constructed to display the composition, resource flows and information flows of the SoS, which can support decisions.

J. Chen et al. (Eds.): KSS 2022, CCIS 1592, pp. 173–185, 2022.
https://doi.org/10.1007/978-981-19-3610-4_13

Architecture models are generally constructed by professional modelers. They construct architecture models in appropriate forms to describe SoS architecture. Due to the design and modeling of SoS architecture are "data-centric", the collection of architecture data is particularly important in the modeling process [2]. However, most SoSs are highly complex. Architecture design requires professional knowledge in multiple fields and disciplines, which leads to the fact that modelers cannot accurately grasp all architecture data. At the same time, architecture data mostly exist in the form of architecture description texts and do not form standard databases or models. With the explosive growth of architecture complexity, the traditional method of manually collecting data will be inefficient and a waste of resources. In addition, another shortcoming exposed in architecture modeling is the shortage of professional modelers. Architecture modeling is mainly completed by modelers. They must understand the decision-making requirements and the specification of architecture modeling. But training modelers need a certain period. Those modelers without professional education cannot complete this task well.

Therefore, when disadvantages of the manual architecture modeling method are exposed, the existing knowledge extraction technologies and model transformation technologies can be considered to intelligently generate architecture models. Computers complete the data acquisition and model generation instead of staff. The intelligent modeling framework of SoS architecture promotes the development of architecture modeling towards intelligence.

2 Research Background

2.1 Construction of Knowledge Graph

A knowledge graph contains a large number of entities, concepts, attributes, and their semantic relationships. In essence, it is a kind of semantic network [3]. Compared with text and tables, graph databases can reveal the relationships between entities more intuitively. Knowledge graph construction technology mainly includes knowledge extraction, knowledge fusion, and knowledge processing [4, 5]. Knowledge extraction, the key to knowledge graph construction, identifies and extracts basic knowledge from multi-source data. Knowledge extraction generally includes named entity recognition [6, 7], relationship extraction [8], and attribute extraction. Entities can be extracted based on rule definition, statistical model, or neural network [9]. Relationship extraction is to extract the relationships between entities. Relationships can be extracted based on rule definition, and neural network. Attribute extraction can be regarded as a kind of relationship extraction.

Storing knowledge based on graph databases to form knowledge graphs is the most popular way of knowledge storage at present. Neo4j is the most popular high-performance NoSQL graph database, which is composed of nodes, edges, and their attributes [10]. In addition, knowledge modeling tools, including Protégé, CiteSpace, Pajek, and so on, can also be used to directly construct the knowledge graph. Among them, Protégé supports the construction of ontology to better represent the logical relationship between ontology and instance data. And this tool has the functions of logical checking, knowledge query, and knowledge reasoning. CiteSpace, the most widely used knowledge graph software in China, is specially used for the construction of scientific knowledge graphs. Pajek is a social network analysis software to support large-scale network decomposition and network relationship display. After constructing the SoS architecture knowledge graph, it is not difficult to find that the knowledge graph has a network structure and multiple types of nodes. It essentially belongs to the category of heterogeneous networks. At present, some researchers have tried to combine knowledge graphs and heterogeneous networks. Due to the rich semantic relationships of knowledge graphs, the integration of the two still has challenges.

2.2 SoS Architecture Modeling

At present, the SoS architecture models are mainly constructed manually by modelers. They should first clarify user requirements and the scope of architecture, then collect relevant architecture data and construct applicable architecture models. These architecture models generally follow the specifications of an architecture framework. A variety of architecture frameworks have been proposed successively, such as the US Department of Defense Architecture Framework (DoDAF) [11, 12], the UK Department of Defense Architecture Framework (MoDAF) [13], and the Unified Architecture Framework (UAF) [14], in which DoDAF is the most widely used. On the other hand, there are a variety of architecture modeling tools, such as IBM Rational Rhapsody, Easy Model, and Special Model, but these modeling tools only make the elements in the models more standardized and the modeling process smoother. The construction of architecture models is still done manually by modelers. When the SoS is too complex, architecture modeling will still consume a lot of resources.

In the field of SoS architecture research, there are no relevant researches and methods of architecture intelligent modeling. In order to solve various problems of traditional architecture modeling methods, this paper creatively proposes an architecture intelligent modeling framework. The framework selects knowledge graph construction technology and model transformation as method support to realize the intelligent construction of architecture models. Automatic data collection is the core step of architecture intelligent modeling. At present, named entity recognition and relationship extraction methods in knowledge graph construction technology are the most popular and effective methods to extract data triples from text data. Moreover, the knowledge graph can clearly present the network relationships and semantic attributes of architecture data, which can support the mining of hidden architecture data. Therefore, the framework proposed in this paper selects knowledge graph construction technology to automatically collect and mine architecture data. Because the architecture models should be standard and shareable. the framework constructs the data-model transformation rules based on the guidance of the architecture framework to generate the architecture models that meet the requirements of users.

3 Framework of SoS Architecture Intelligent Modeling

The SoS architecture intelligent modeling takes the generation of architecture models as the purpose. The overview of this framework is shown in Fig. 1. A large amount of architecture data exist in the architecture description texts. Therefore, these description texts are used as the original data. Firstly, architecture data need to be extracted from the architecture description texts. The extraction results are knowledge triples in the forms of "entity-relationship-entity" and "entity-attribute-attribute value". And the architecture knowledge graph is constructed based on these knowledge triples. Due to the imperfect description text and the error of knowledge extraction, the extracted architecture data is not complete. Therefore, the second step needs to complete the architecture knowledge based on the extracted data. Finally, the data-model transformation rules are formulated to automatically generate architecture models.

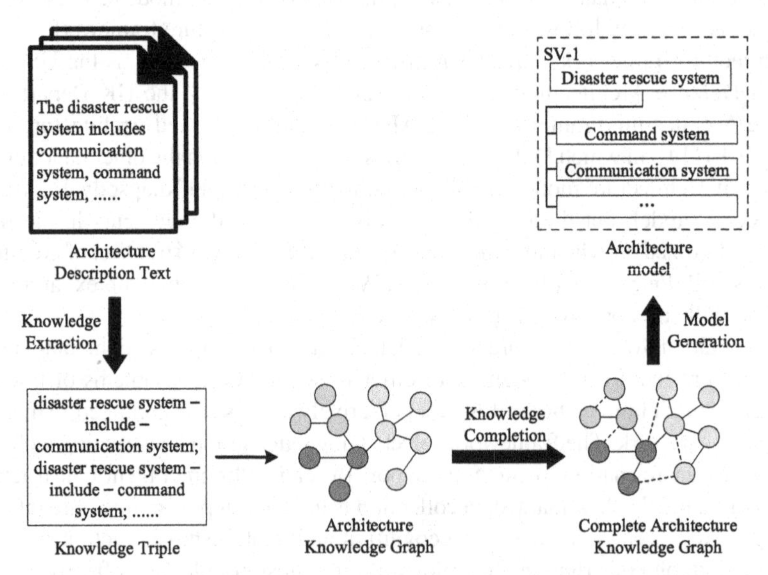

Fig. 1. Overview of SoS architecture intelligent modeling framework

Based on the overview, this paper proposes a framework of architecture intelligent modeling based on knowledge graph, as shown in Fig. 2. The framework, including three modules: knowledge extraction, knowledge completion, and model generation, aims to realize the automatic collection of architecture data and the intelligent construction of the architecture models.

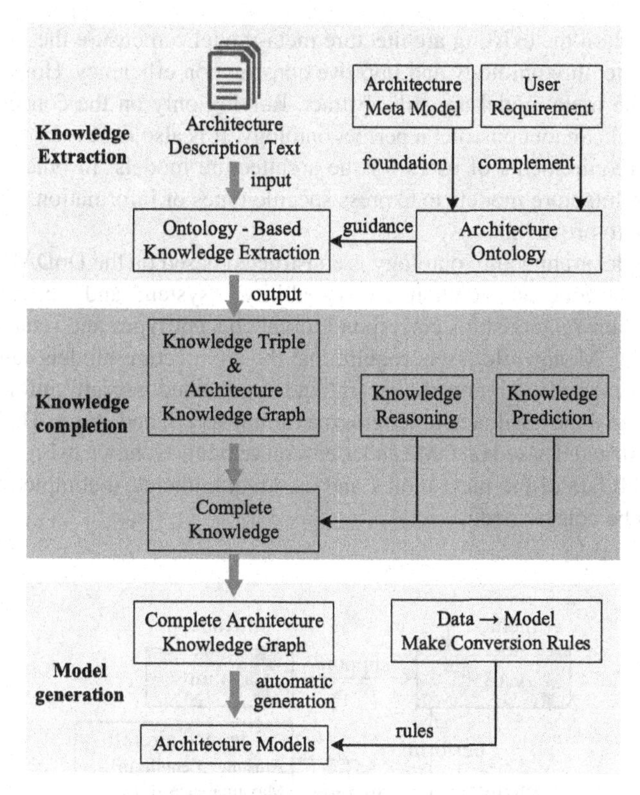

Fig. 2. Intelligent modeling framework for SoS architecture based on knowledge graph

3.1 Architecture Knowledge Extraction

Data is the basis of architecture model construction. Based on knowledge extraction technologies, existing standardized data from architecture description texts is the key process of architecture intelligent modeling. In the framework designed in this paper, knowledge extraction includes architecture ontology construction, architecture data extraction, and architecture knowledge graph construction.

Architecture Ontology Construction. In the field of architecture modeling, there are high requirements for the standardization of architecture data. Architecture data must be able to support resource sharing and data reuse. The premise of constructing an architecture knowledge graph is to construct architecture ontology. As the model layer of the knowledge graph, ontology is the logical basis of the knowledge graph. Ontology defines concept types, attribute types, and relationship types. Architecture ontology clarifies the data types that need to be extracted from the architecture description texts. Meanwhile, it standardizes the architecture data obtained by knowledge extraction to ensure that these data meet the modeling requirements. The construction of ontology requires professional knowledge to ensure the standardization and integrity of ontology. Therefore, it is very difficult and a waste of resources to construct ontology directly. Constructing ontology based on the existing architecture meta-model can ensure the logical specification of architecture ontology and improve construction efficiency. However, the data concepts in the meta-model are still abstract. Relying only on the conceptual data in the meta-model cannot construct a perfect ontology. It is also necessary to consider the construction requirements of users for the architecture models. In other words, users require the architecture models to express specific types of information. The following is an example to prove this.

For example, architecture ontology is constructed based on the DoDAF meta-model (DM2) [15]. The conceptual data such as "capability", "system" and "activity" contained in DM2 and their relationships correspond to the concept types and relationship types in the ontology. Meanwhile, users require that the architecture models can present not only "system", but also "manned aircraft" and "unmanned aircraft" information. And they need to have the attributes of length, maximum speed, and so on. The architecture ontology constructed based on DM2 and user requirements is shown in Fig. 3. Integrating the conceptual data of the meta-model and user requirements, a complete architecture ontology can be constructed.

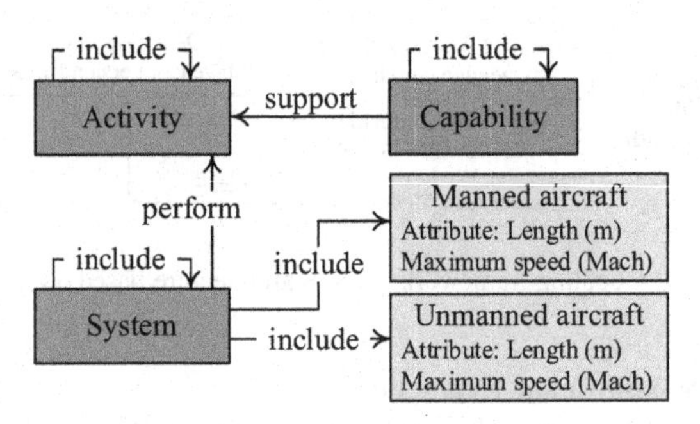

Fig. 3. Sample of architecture ontology

Architecture Knowledge Extraction. After constructing architecture ontology, the data types that need to be extracted are defined. Taking the architecture description texts as the input, the knowledge extraction process collects architecture data in the form of knowledge triples.

The first step of knowledge extraction is named entity recognition. Its purpose is to extract named entities such as "capability", "system", and "activity" from source data. The entity data will be used as the input of relationship extraction. So, the effect of named entity recognition has a great impact on the accuracy of architecture data. At present, named entity recognition can be completed by rule-based, statistical model-based, and neural network-based methods. The rule-based method requires experts to make rules manually, but it has good results in some professional fields. The statistical model-based method, including Hidden Markov Model (HMM), Conditional Random Fields (CRF), and so on, are trained with manually labeled corpus. The neural network-based method is the most popular method at present, such as Bidirectional Long Short-Term Memory

(Bi-LSTM) [16], BERT training language model [17], and so on. Part of the entity data is shown in Fig. 4.

ACT:["air defense and antimissile" ,
"reconnaissance mission" ,
"fire strike" , ···]
CAP:["reconnaissance capability" ,
"fire strike capability" , ···]
SYS:["early warning satellite" ,
"F-15" , ···]

Fig. 4. Part of architecture named entity recognition results

Based on entity data, the relationships between entities are extracted. Due to the rich semantic relationships of the SoS architecture, relationship extraction is more complex than entity recognition. Relationship extraction can be completed by rule-based or neural network-based methods. Relying on manually defined extraction rules, the rule-based method is relatively simple. The neural network-based method is complex, which is the research focus in the field of knowledge extraction. Part of the relationship data is shown in Table 1.

Table 1. Part of relationship extraction results

Head entity	Tail entity/Attribute value	Relationship/Attribute
Reconnaissance capability	Reconnaissance mission	Capability-support-activity
Early warning satellite	Reconnaissance capability	System-have-capability
F-15	Fire strike	System-perform-activity
F-15	2.5 Mach	Maximum speed
...	...	...

Architecture Knowledge Graph Construction. After getting the architecture data, storing the data in the form of knowledge graph can make the data expression more intuitive. Meanwhile, the network structure of the data association is highlighted.

There are many ways to construct knowledge graphs. Among them, Neo4j is the most widely used graph database. When using Neo4j to construct a knowledge graph, the third-party database should be installed in advance, which is of great help to quickly construct the knowledge graph. It must be noted that Neo4j currently only supports

Java and Python languages. In addition, existing knowledge modeling tools can also be used to construct knowledge graphs, such as Cite Space, Protégé [18], etc. Part of the architecture knowledge graph is shown in Fig. 5.

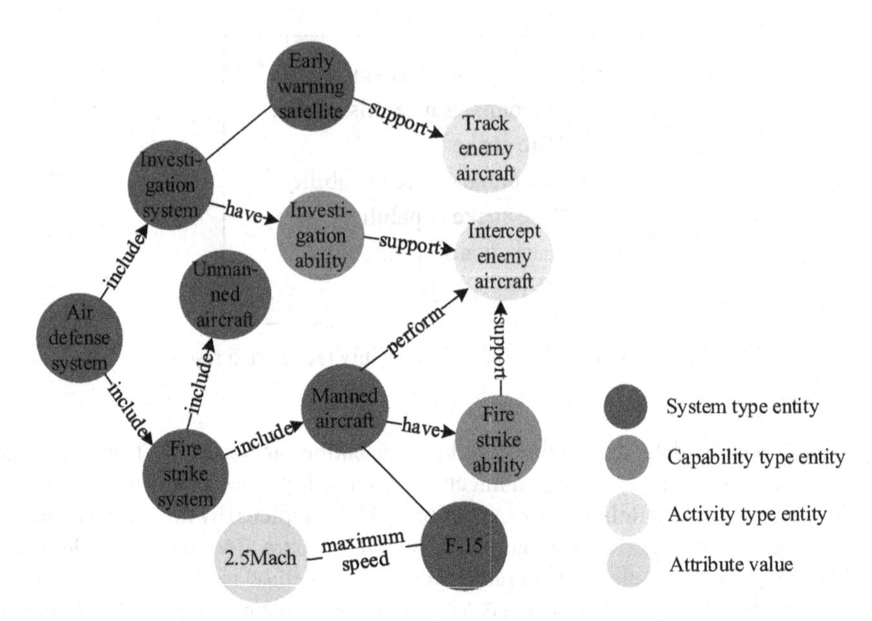

Fig. 5. Part of architecture knowledge graph

3.2 Architecture Knowledge Completion

Due to the incompleteness of architecture description text and the error of knowledge extraction, there is a lack of data in the architecture knowledge graph. These missing data will affect the subsequent model construction, so it is of great significance to complete the architecture knowledge graph. After analyzing the entity types and relationship types of the architecture knowledge graph, it is found that most of the missing data can be obtained by knowledge reasoning. For the missing knowledge that cannot be obtained by knowledge reasoning, the methods of heterogeneous network link prediction are migrated to complete the knowledge.

Knowledge Reasoning. Knowledge reasoning is a common data mining method in the field of knowledge graph construction, which mines the implicit semantic relationships according to the existing facts. In the architecture knowledge graph, the reasoning of entity relationship is the focus. By analyzing various semantic relationships, the logical reasoning path is formulated as reasoning rules. The knowledge reasoning path that can infer the target knowledge can be determined from the ontology.

There are many ways to reason knowledge. The most simple and effective way is to use the reasoning engine in the knowledge modeling tools. For example, the knowledge modeling tool Protégé has its Pellet inference engine. It can complete knowledge reasoning quickly only based on defined inference rules. In the air defense SoS, if a system can intercept enemy aircraft, the system must have firepower strike capability. In other words, from the two relationships of "system-perform-enemy aircraft interception" and "fire strike capability-support-enemy aircraft interception", "system-has-fire strike capability" can be inferred. The reasoning rule is shown in Fig. 6.

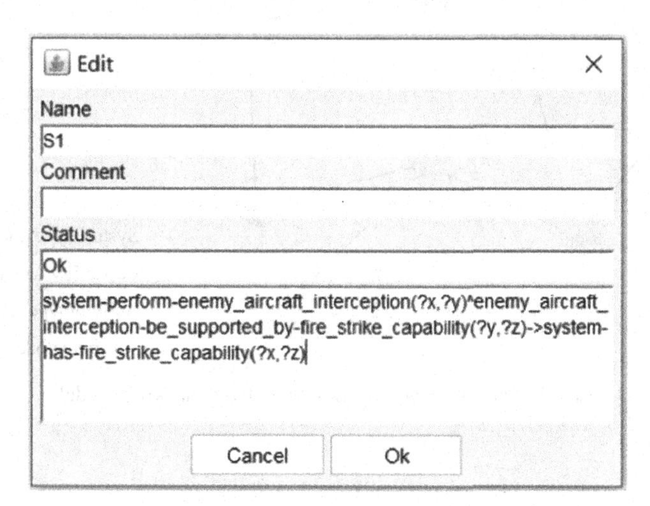

Fig. 6. Example of reasoning rules definition

Knowledge Prediction. Some types of implicit knowledge in the architecture knowledge graph do not have a clear reasoning path. So, these types of knowledge cannot be obtained by knowledge reasoning. For example, through the path "fighter plane-have-fire strike capability-support-enemy aircraft interception", "fighter plane-execute-enemy aircraft interception" can be inferred. However, through the path "early warning satellite -have- reconnaissance capability-support -enemy aircraft interception", "early warning satellite - execute - enemy aircraft interception" cannot be inferred. Knowledge reasoning cannot realize the mining of similar types of implicit knowledge.

Being similar to heterogeneous networks, the architecture knowledge graphs have obvious network structure. The entities and relationships in the architecture knowledge graph can be modeled as a knowledge network, as shown in Fig. 7. In many papers, the relationship between heterogeneous networks and knowledge graphs is analyzed. Knowledge graph essentially belongs to the category of heterogeneous network. So, the models and algorithms in the research of heterogeneous networks can be applied to the analysis of knowledge graphs. Therefore, this paper considers migrating heterogeneous network link prediction for knowledge prediction. The semantic relationship and network structure of the knowledge graph are comprehensively considered to predict the probability of relationships between entities.

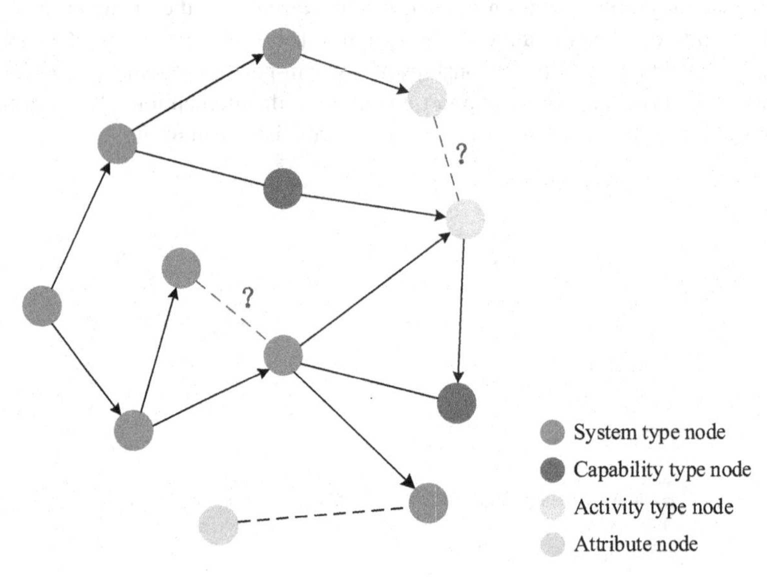

Fig. 7. Part of the architecture knowledge network model

3.3 Architecture Model Generation

After the above steps, the architecture data has been extracted from the architecture description texts. And the relevant data are stored in the architecture knowledge graph. The ultimate goal of the SoS architecture intelligent modeling framework is to generate architecture models. Therefore, the last step is formulating data-model conversion rules to realize the automatic generation of architecture models.

Architecture Model Classification. Firstly, the categories of architecture models need to be sorted and classified. There are many kinds of architecture models. These models include different types depending on the presentation form, such as table type, structure type, behavior type, mapping matrix type, ontology type, schedule type, and so on. The transformation rules formulated to generate the same type of models are similar. So, determining the type of architecture model and the required architecture data in advance can formulate data-model transformation rules more efficiently. Taking DoDAF2.0 as an example, the types and required data of some models are shown in Table 2.

Table 2. Types and data requirements of some models

Architecture model	Required data types
Structure type	
Capability dependencies (CV-4)	Capability
	Capability-include-capability
	Capability-support-capability
	Capability-synergy-capability
Mapping matrix type	
Capability to operational activities mapping matrix (CV-6)	Capability
	Activity
	Capability-support-activity

Formulation of Data-Model Transformation Rules. After clarifying the types and required data of architecture models, the models are automatically generated by defining the transformation rules between data and models. For example, the generation of the system interface description model (SV-1) requires "system" type data. The generation rules of the SV-1 model are as follows:

Rule 1: Each instance of the "system" type entity generates a block module with the same name in the model.
Rule 2: If there is a relationship between two "system" type entities, the association line is used to link the two entities. And the relationship type is named as the label of the corresponding association line.
Rule 3: If the "system" type entities have attributes, they are added to the new attributes of the block module.

Based on the generation rules constructed above, the SV-1 model is generated from the five data triples of "fire strike system - include - UAV", "fire strike system - include - MAV", "UAV - coordination - MAV", "MAV - include - F-15", "F-15 - maximum speed – 2.5 Mach", as shown in Fig. 8.

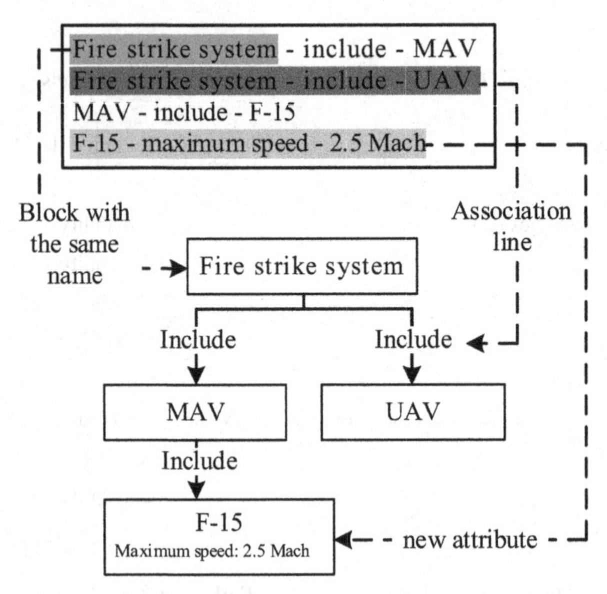

Fig. 8. System interface description model generation

4 Conclusion

The intelligent modeling framework of SoS architecture proposed in this paper includes knowledge extraction, knowledge completion, and model generation. The framework integrates cutting-edge data processing technologies such as natural language processing and knowledge reasoning. Architecture description texts are taken as the input data to automatically generate architecture models. Solving the disadvantages of the traditional manual method of architecture modeling, the framework promotes the development of architecture modeling towards intelligence.

At present, this paper has completed the process design of SoS architecture intelligent modeling. In the follow-up work, the specific implementation methods will be researched. After comparative experiments, the optimal implementation method is selected to form a complete SoS architecture intelligent modeling method.

References

1. Eung-jo, D.: Race to the algorithmic warfare: an analysis of the China's system confrontation and the USA's systems warfare. J. Strate. Stud. **28**(3), 217–264 (2021)
2. Prasanth, C., et al.: Marvel: a data-centric approach for mapping deep learning operators on spatial accelerators. J. ACM Trans. Archit. Code Optim. **19**(1), 1–26 (2022). https://doi.org/10.1145/3485137
3. Hogan, A., et al.: Knowledge graphs. ACM Comput. Surv. **54**(4), 1–37 (2021)
4. Xu, Z., Sheng, Y., He, L., Wang, Y.: Review on knowledge graph techniques. J. Univ. Electron. Sci. Technol. China **45**(4), 589–606 (2016)
5. Liu, Q., Li, Y., Duan, H., Liu, Y., Qin, Z.: Knowledge graph construction techniques. J. Comput. Res. Dev. **53**(3), 582–600 (2016)

6. Jiao, K., Li, X., Zhu, R.: Overview of Chinese domain named entity recognition. Comput. Eng. Appl. **57**(16), 1–15 (2021)
7. Zara, N., Syed, W., Muhammad, K.: Named entity recognition and relation extraction: State-of-the-Art. ACM Comput. Surv. **54**(1), 1–39 (2021)
8. Haihong, E., et al.: Survey of entity relationship extraction based on deep learning. J. Softw. **30**(6), 1793–1818 (2019)
9. Deng, Y., Wu, C., Wei, Y., Wan, Z., Huang, Z.: A survey on named entity recognition based on deep learning. J. Chin. Inf. Process. **35**(9), 30–45 (2021)
10. Naga, D.: Book recommendation system using Neo4j graph database. Int. J. Anal. Exp. Modal Anal. **12**(6), 498–504 (2020)
11. The DoDAF Architecture Framework Version 2.02. https://dodcio.defense.gov/Library/DoD-Architecture-Framework/
12. Wang, Y., Bi, W., Zhang, A., Zhan, C.: DoDAF-based civil aircraft MBSE development method. Syst. Eng. Electron. **43**(12), 3579–3585 (2021)
13. Ministry of Defense Architectural Framework (MODAF). https://wenku.baidu.com/view/a9cc3c3a376baf1ffc4fad06.html
14. Ding, Q., Wang, Y., Shen, Y., Ge, L.: Methodological significance of UAF. J. Mil. Oper. Res. Syst. Eng. **32**(4), 63–67 (2018)
15. Holly, A.: Incorporating the NATO human view in the DoDAF 2.0 Meta model. Syst. Eng. **15**(1), 108–117 (2012)
16. Yan, R., Jiang, X., Dang, D.: Named entity recognition by using XLNet-BiLSTM-CRF. Neural Process. Lett. **53**(5), 3339–3356 (2021). https://doi.org/10.1007/s11063-021-10547-1
17. Sun, C., et al.: Biomedical named entity recognition using BERT in the machine reading comprehension framework. J. Biomed. Inform. **118**, 103799 (2021)
18. Manoj, K., Tanveer, J.: An ontology construction approach for retrieval of the museum artifacts using Protégé. Int. J. Comput. Sci. Issues **13**(4), 47–51 (2016)

Introducing Trigger Evolutionary Graph and Event Segment for Event Prediction

Yaru Zhang[1,2(✉)] and Xijin Tang[1,2]

[1] Academy of Mathematics and Systems Science, Chinese Academy of Sciences, Beijing 100190, China
{zhangyaru,xjtang}@amss.ac.cn

[2] University of Chinese Academy of Sciences, Beijing 100049, China

Abstract. Event chain is composed of a sequence of events which are closely related. Event prediction models aim to predict the most possible following event given the existing chain. Previous approaches based on event pairs or the whole chain may ignore much structural and semantic information. Current datasets for event prediction, naturally, can be used for supervised learning. Chains are either from document-level procedural action flow, or from news sequences under the same column. This paper leverages graph structure knowledge of event triggers and event segment information for event prediction with general news corpus, and adopts standard multiple choice narrative cloze task evaluation. The topic analysis method is used to extract event chains from the news corpus. Based on trigger-guided structural relations in the event chains, we construct trigger evolutionary graph, and trigger representations are learned via Graph Convolution Neural Network. Then there are features of two levels for each event, namely text level semantic feature and trigger level structural feature. We learn the features of event segments split according to event subjects by attention mechanism, and integrate relevance between the candidate event and each event segment. The most possible following event is selected by the relevance. Experimental results on the real-world news corpus demonstrate the effectiveness of the proposed model.

Keywords: Event prediction · Trigger evolutionary graph · Event segment

1 Introduction

An event is a specific thing which happens at a particular time and place [2]. Typically, there are two ways to present event. One is narrative, in rich text, such as news report. The other is formalized tuple composed of event type, time, place,

Supported by National Natural Science Foundation of China under Grant Nos. 71731002 and 71971190.

J. Chen et al. (Eds.): KSS 2022, CCIS 1592, pp. 186–201, 2022.
https://doi.org/10.1007/978-981-19-3610-4_14

etc. key elements. Events often occur in a sequence, and the sequence of events is called event chain. Given text description or procedural action flow, understanding events automatically is of practical significance. Therefore, recent years witness many studies about events, such as event detection [13], event extraction [11,21,24], storyline extraction [3,19]. Event detection aims to identify event trigger type. By means of event extraction methods, we could formalize the event as the event triggers and corresponding event arguments. Storyline extraction is used to obtain event evolutionary process, where event pair relations, such as temporal relation and causal relation [8,16] need to be determined. Compared with these tasks, event prediction is a more challenging task, and could be seen as an upstream task for them. Given the existing event chain, event prediction requires models to predict the most possible following event. According to task types, there are three kinds of event prediction tasks. The first one is classification and regression task, which focuses on predicting the type and timestamp of the next event [1,20]. Of course, there are some researches which only handle event type prediction problem [5,12]. The second one is standard multiple choice narrative cloze task. Given existing event chain and candidate event set, generally, which contains the real next event and some negative samples, it is expected that the model could select the real next event. The last one is natural language generation task. Given history events, the model is expected to generate text description of the next event. Current methods mainly operate on LSTM encoder and LSTM decoder [6,7,18].

This paper pays attention to the second kind of event prediction task. Event pair based method which considers association between each history event and the candidate event to select the next event is one basic method for this task [4]. However, event pair level operation does not capture much semantic information within the event chain. Wang et al. propose a dynamic memory network model, which combines the advantages of both LSTM temporal order learning and traditional event pair coherence learning to infer the next event [22]. Some researchers construct event evolutionary graphs to learn abstract event evolutionary principles and patterns for event prediction [10,15,25]. Modeling the whole chain may incur redundant information. Lv et al. conduct an event segment level study [14]. They utilize self attention mechanism to capture diverse event segment information within the chain. Event level attention is used to model event pair relations. Chain level attention is used to model the relations between the next event and self attention history events. This work learns event segment information by self attention of event chain. In fact, the information is still of chain level focusing on different individual events. Totally, above methods are usually used to process standardized fine grained events, often with one common subject in each chain. Event chain composed of fine grained events is of short dependence, which makes prediction easier. There exists event prediction under more complex scenarios. As we further indicate, there are no natural event chains, and event prediction model needs to fully understand strategies of different agents in the manually assembled event chain involving multiple agents, then predicts the next event.

Thus, it is more difficult to understand semantics and perform inference in this scenario.

In view of the fact that both subjects and triggers are key elements of events, here, we define event prediction as a standard multiple choice narrative cloze task, and leverage graph structure knowledge of event triggers and semantics of event segments generated according to event subjects for event prediction on the real-world news corpus. Firstly, considering event keywords and entities, we use the topic analysis method to extract event chains from the news corpus. Then, inspired by the conversation graph research [23], where transition knowledge about keywords is used to generate response, we construct the event trigger evolutionary graph, and learn trigger representations by means of Graph Convolution Neural Network (GCN) and the transformation relations based on triggers between the next event and the existing chain. An event is represented as the concatenation of text level feature and trigger level feature. Afterwards, our model learns the feature of event segment i.e. subsequence of the chain to integrate the sequence information of the same subject. Finally, the score of the candidate event is determined by relevance between event segments and the candidate event. We summarize our main contributions as follows:

- We construct event trigger evolutionary graph, and learn trigger-guided transition patterns among events for helping predicting the next possible action.
- We learn hierarchical features of different event segments generated according to event subjects, and model relations between the candidate event and these event segments.
- Different from existing event prediction circumstances, we conduct experiments on the general news corpus to show the performance of the proposed model.

2 Related Work

According to task types, there are three main event prediction tasks, classification task, ranking task (or multiple choice cloze task) and natural language generation task. Only when the type of the next event needs to be determined, is the problem defined as a classification task. Taymouri et al. adjust generative adversarial networks (GAN) to sequential temporal data for the prediction of the next event label and timestamp [20]. Heinrich et al. introduce gated convolutional neural networks (GCNN) and key-value-predict attention networks (KVP) to learn chain level feature for the next event label inference [5]. With the development of knowledge graph, Peng et al. develop pairwise popularity graph convolutional network (PP-GCN) to identify event categorization [17]. Knowledge graph is also applied to event prediction to enrich the representations of events with entities and relations [9,12]. With creation of paths and schemas, modelling methods and evaluation metrics become richer. Not only can [9] deal with event type prediction problem, but also can do schema matching. Nevertheless, extra knowledge needs to be got in advance.

When an event needs to be described with more elements besides type, future events may become more diverse. In this case, there are two kinds of ways to process event prediction problem. One way is to provide candidate event set for given history events, then the prediction model selects the most appropriate next event from among, or gives the ranking of these candidate events. The way is similar to classification task about finite event types, and tailored candidate is just set for each existing chain. The other is to generate the text description of the next event. There are many studies which handle event prediction problem given candidate event sets. Granroth-Wilding et al. propose a compositional neural network model which composes embeddings of words describing events into event representations, and then combines the strength of association between each history event and the candidate event to select the next event [4]. Nevertheless, event pair level operation does not capture rich semantic feature within the event chain. Wang et al. model temporal relationship of the chain using LSTM, and then adopt a dynamic memory network for relation measuring [22]. Li et al. construct an event graph, and use gated graph neural network (GGNN) to model event evolutionary information for inferring the next event [10]. There are other event evolutionary graph based approaches [15,25] for event prediction, similarly, which focus on learning abstract event evolutionary patterns. Modeling the whole chain may bring irrelevant noise. We will learn the transition relationship among event triggers to help get action aspect association between the next event and existing chain. Lv et al. suppose that event segment information facilitates the prediction of the next event [14]. They develop self attention mechanism to learn event segment information within the chain. However, the information is still chain level information focusing on different individual events. This paper will learn the features of event segments from event subject perspective to integrate semantics of the same agent.

The next event text generation task is often for short term event chain, where the context is very closely related. Typically, Hu et al. utilize LSTM to encode existing event chain, then adopt LSTM decoder to generate the text description of the next event [6]. Hierarchical attention mechanism and semantic encoder are used to improve the performance of the model [7,18].

3 Proposed Model

In this section, we will introduce event trigger evolutionary graph and event segment for event prediction. The overall structure of our model is shown in Fig. 1. Firstly, we present the extraction method of event chains, and construct the candidate set for each event chain sample. The second step is to learn text level representations of events. The third step is to construct event trigger evolutionary graph, and learn event trigger representations. Then we split event chains into event segments according to event subjects, and obtain the feature of each event segment by the attention mechanism. Finally, relevance between the candidate event and each event segment is integrated to get the score of the candidate event.

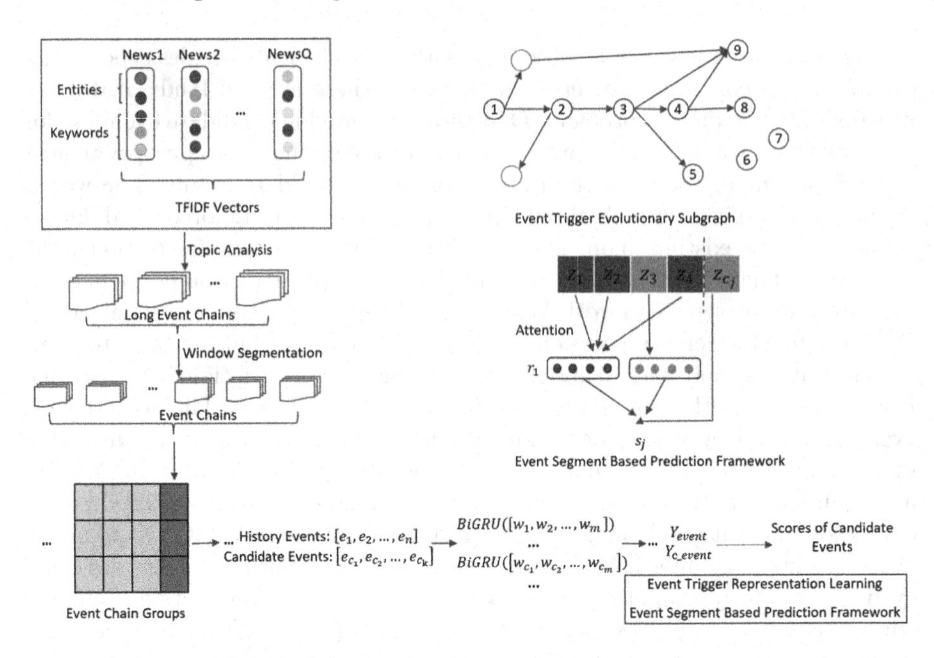

Fig. 1. Overall framework of event prediction.

3.1 Extraction of Event Chains

Since there are not well arranged event chains in the general news corpus, this paper needs to assemble event chains according to temporal order and semantic association in order to carry out event prediction research. The quality of the event chains may influence the result of event prediction. Concretely, we utilize stanza[1] developed by Stanford University to perform Named Entity Recognition (NER) and Part-Of-Speech (POS) tagging on the news content. Specific types of entities, {'PERSON','ORG','GPE','LOC','PRODUCT','MONEY'}, namely, {'person', 'organization', 'geo-political entity', 'location', 'weaponry', 'money'}, are selected as the entity characteristics of the news. Specific part-of-speech words, {'NOUN','VERB','ADJ'}, namely, {'noun', 'verb', 'adjective'}, are selected and serve as the keyword characteristics together with the given keywords of the news.

Afterwards, for entities and keywords, we respectively select high-frequency words (word frequency is greater than or equal to 15) to construct two kinds of TFIDF vectors for each news sample, and keywords repeated with entities will not be considered. Based on concatenation of two vectors, $d_e \oplus d_k$, we carry out LDA[2] topic analysis on the news corpus. Considering expected average event chain length, we set the number of topics. After then, the topic label of each news sample is obtained.

[1] https://github.com/stanfordnlp/stanza.

[2] https://blog.csdn.net/huagong_adu/article/details/7937616.

We arrange the news samples with the same topic according to the news release time, so as to derive the long event chains. Each event is represented by corresponding news title. Too long event chain may introduce noise, which interferes event prediction. Therefore, we divide the long event chains according to a certain window size into the non-overlapping short event chains to get the event chain sample set.

In this study, event prediction is defined as a multiple choice cloze task. Each time, k event chains are extracted from the event chain set without repetition, and their last events form the shared candidate event set. So for each existing chain, there are 1 real following event and $k-1$ negative samples in the candidate set. Two ways, random extraction and sequential extraction, are utilized. Random extraction will make the topics of events in the candidate set tend to be diverse. In order to enable the model to select the real next event from the candidate set with high confusion, we apply sequential extraction. The topics of events in the candidate set are usually consistent in this way, which is more matched with the real scenario. The ratio of the number of the samples from two extraction ways is 2:1.

3.2 Event Representation Learning

Each word in the news title is initially embedded based on Glove[3]. After word embedding, the event e is regarded as a sequence of word vectors: $e = [w_1, w_2, \ldots, w_m]$, where m is the length of the sequence. Then, we encode the word based sequence by BiGRU which contains a forward $\overrightarrow{GRU}$ and a backward $\overleftarrow{GRU}$,

$$\overrightarrow{y_t}, \overrightarrow{h_t} = \overrightarrow{GRU}(w_t, \overrightarrow{h_{t-1}}) \tag{1}$$

$$\overleftarrow{y_t}, \overleftarrow{h_t} = \overleftarrow{GRU}(w_t, \overleftarrow{h_{t-1}}) \tag{2}$$

where $\overrightarrow{h_t}, \overleftarrow{h_t}$ represent hidden states of both directions at time t, $\overrightarrow{y_t}, \overleftarrow{y_t}$ represent outputs of both directions at time t. We treat y, $y = \overrightarrow{y_m} \oplus \overleftarrow{y_1}$, as the semantic representation of the event.

3.3 Trigger Representation Learning

The evolution of events, to some extent, can be regarded as the transition among event triggers. This subsection learns the representations of event triggers which will serve as the second level features of events besides semantic features. Compared to event chains with standardized fine grained events, the triggers in the event chains based on news titles lack standardization, and their description styles are more diverse. Therefore, based on the transition relationship among triggers this paper will adopt GCN to integrate neighborhood information, and further get the representations of triggers.

[3] https://github.com/stanfordnlp/GloVe.

In particular, we construct the event trigger evolutionary graph G according to event chains in the training set. The graph G is a weighted directed graph, where the nodes are triggers, and edges denote transition times between two triggers. The initial feature of the node is the corresponding Glove embedding. The representation of the central node is updated through the following convolution operation,

$$x_i^{(0)} = ReLU(W_o w_i + b_o) \tag{3}$$

$$x_i^{(l)} = W_g^{(l)}(x_i^{(l-1)} + \sum_{j \in N_i} w_{ij} x_j^{(l-1)}) + b_g^{(l)} \tag{4}$$

The initial feature of the node w_i is projected by parameters W_o and b_o. N_i denotes neighborhood set of the central node. w_{ij} which is the weight between the neighbor node and the central node is normalized according to all neighbors of the central node. Learnable parameters $W_g^{(l)}$ and $b_g^{(l)}$ are used to update the aggregation result, and then we obtain the representation of the central node $x_i^{(l)}$ after l convolution operations (one convolution operation is used here).

In order to contact triggers of history events with triggers of candidate events, we define different types of neighbors for them. Then the above GCN is used to aggregate information of corresponding neighbors, and representations of these event triggers are updated. Figure 2 is a trigger evolutionary subgraph including history event trigger words 1–4 and candidate event triggers 5–9. For the triggers in the first half of the existing event chain, we select their own out-degree neighbors. In other words, we focus on what triggers they will point to. Considering media effect of triggers in the second half of the existing events, we use all neighbors to update their features. The triggers in candidate events are updated using their own in-degree neighbors, namely, we focus on what triggers will point to them. Based on these pointing and being pointed relations, node 9 in Fig. 2 is directly related to node 3 and node 4, and indirectly related to node 1. The event containing the trigger is very likely to be the real next event. All in all, we expect to use the transformation relationship among triggers to mine the correlation among events from the perspective of action transformation.

3.4 Event Segment Representation Learning

Individual event in the existing event chain is related to the next event to be predicted, and semantics of the whole event chain also plays an important role in prediction. As mentioned, just considering event pair or the whole event chain may result in information insufficience or information redundancy. The event chain could be divided into event segments of different lengths according to different rules. Event segments integrate semantics of a certain level, which could generate a compromise effect between individual event and event chain. Based on coarse-grained major subjects of events this paper divides event chains to get different event segments. Each event in the chain describes action of a certain major subject. Events with the same major subject form the event segment.

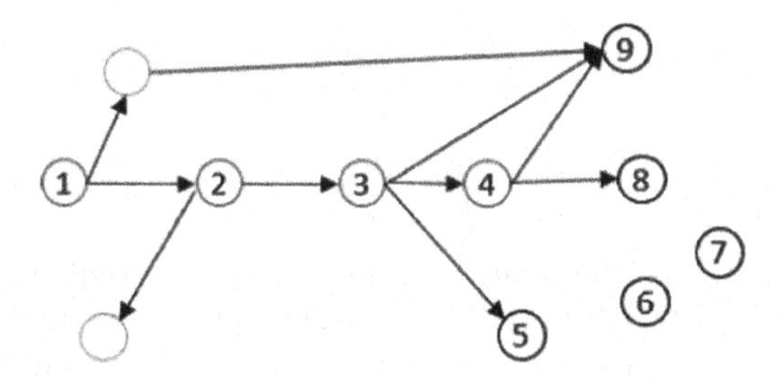

Fig. 2. Event trigger evolutionary subgraph.

As shown in Fig. 3, different colors represent different major subjects, $z_1, ..., z_4$ denote the representations of history events, and z_{c_j} denotes the representation of the candidate event. Thus there are two event segments in the existing chain for this case. Next, we elaborate how to learn the features of event segments and calculate the final score of the candidate event.

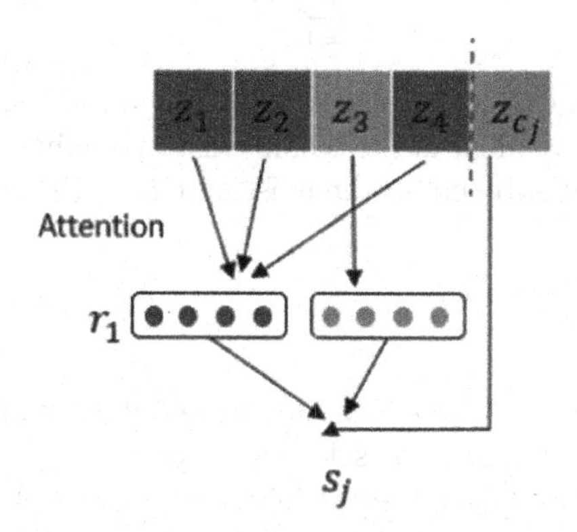

Fig. 3. Event segment based prediction framework.

Specifically, we concatenate the event semantic representation and the event trigger representation to get the event vector $z, z = y \oplus x$. Assuming that an event chain usually includes interaction between two major subjects, we divide the matrix composed of the history event vectors into three submatrixes (i.e. three segments) according to the subjects. They respectively denote the event matrix corresponding to the major subject with the highest frequency, the event matrix corresponding to the major subject with the second highest frequency, and the event matrix corresponding to others.

For the first event segment and the candidate event representation z_{c_j}, we utilize the following "Attention" mechanism to calculate weight of each event in the segment, and further get the representation of the event segment r_1,

$$M_1 = \phi(W^{Z_1}[z_1, z_2, \ldots, z_s] + [W^H z_{c_j}, \ldots, W^H z_{c_j}]) \tag{5}$$

$$\gamma_1 = softmax(w^T M_1) \tag{6}$$

$$r_1 = [z_1, z_2, \ldots, z_s]\gamma_1^T \tag{7}$$

where $[z_1, z_2, \ldots, z_s]$ is the matrix formed by the event vectors belonging to the event segment, W^{Z_1} and W^H denote learnable matrix parameters, $[W^H z_{c_j}, \ldots, W^H z_{c_j}]$ contains s column vectors, ϕ is $Tanh$ function, w is learnable vector parameter, and γ_1 denotes weight of each event in the segment. In the same way, we respectively use W^{Z_2}, W^{Z_3} to map the second event segment and the third event segment, and further get the segment representations, r_2, r_3. If there is only one event in the segment, the event vector is the event segment feature.

We accumulate the similarity between each event segment and the j_{th} candidate event to obtain the candidate event score s_j,

$$s_j = -\sum_{i=1}^{seg} ||r_i - z_{c_j}||_2 \tag{8}$$

Generally, seg is equal to 3. If there is only one major subject in the existing chain, the first event segment is the whole existing chain, and seg is equal to 1. If there are two major subjects in the existing chain, r_3 does not exist, and seg is equal to 2.

3.5 Training

Softmax operation is performed on the scores of k candidates to obtain the probability of each candidate being selected, $[\mathrm{P}(\mathrm{e}_{\mathrm{c}_1}), \mathrm{P}(\mathrm{e}_{\mathrm{c}_2}), ..., \mathrm{P}(\mathrm{e}_{\mathrm{c}_\mathrm{k}})]$. The event with the highest probability is the predicted next event. Here, we could choose to apply the following cross entropy loss function to updating the parameters of the model,

$$L = -\frac{1}{N}\sum_{q=1}^{N} log\mathrm{P}(\mathrm{E}_\mathrm{q}^*) \tag{9}$$

where N denotes the number of event chains, $\mathrm{P}(\mathrm{E}_\mathrm{q}^*)$ represents the predicted probability of the real next event in sample q.

Due to randomness, there are various possible distributions for the types of events in the candidate set. If the candidate set contains a large number of events whose types are different from that of the existing event chain, the model can easily select the correct next event relying on the large semantic difference. On the contrary, if the candidate set contains a large number of events whose types are consistent with that of the existing event chain, the difficulty of prediction increases. Namely, if the model can select the correct one from the candidate set with high degree of confusion, it has strong prediction ability. The latter case is more practical.

Therefore, based on the original loss function, this paper sets different weights for different samples according to event type distribution to address the above problem. To be specific, there are the more events whose types are consistent with that of the existing event chain in the candidate set, the greater the weight of the sample is. We calculate the proportion of events whose types are the same as that of the existing event chain in the candidate set, and perform $softmax$ on the proportion of each sample to get the weight of the sample ρ_q. Then ρ_q is added into the original loss function. The formalized process is as follows,

$$L = -\sum_{q=1}^{N} \rho_q log \mathrm{P}(\mathrm{E}_\mathrm{q}^*) \tag{10}$$

$$\rho_q = softmax(\frac{\sum_{j=1}^{k} I(topic_{c_j}^q = topic^q)}{k}), q = 1, 2, \ldots, N \tag{11}$$

where $I(topic_{c_j}^q = topic^q)$ is the indicator function. If the j_{th} candidate event is of the same topic as the q_{th} history event chain, take 1, otherwise take 0.

4 Experiments

4.1 Dataset

This paper conducts event prediction on the Sina Military News corpus for practical experiments. Sina Military News is a portal of Sina News, which contains public reports related to military. We choose the news from 2015 to 2017, and set the number of topics as 1,000 to identify the topic of each news item.

We use the ltp[4] developed by Harbin Institute of Technology to perform Semantic Role Labeling (SRL) on the news titles. Thereby {subject, trigger, object} of each event is identified, and we also obtain word segmentation of each news title. Drawing support from NER and external knowledge base[5], we map the subject to the corresponding major subject. If there are multiple tuples in the news title, generally, we select the first tuple, and the trigger in this tuple is regarded as the event trigger. However, if the trigger is one of "say, talk, comment, intend, want, order and make", the second tuple is selected. There are 9485 words whose frequency are greater than or equal to 5. We take them as the dictionary and use Glove to learn word vectors.

After removing the news with low frequency triggers (i.e. triggers which do not appear in the dictionary) and the news without triggers, there are a total of 37,160 news items. We try to set the length of event chain as 5. However, progressive succession among events is not obvious, so the prediction result is not ideal. This may be due to interference of news timeliness. Considering the long and short semantics distance problem of event chain, we set the length of event chain to 10 in this study, and a total of 3,281 event chains are obtained

[4] http://ltp.ai/.
[5] http://kw.fudan.edu.cn/cndbpedia/.

(short chains with length less than 10 are removed). The number of candidate events k is set to 5, and candidate event sets for event chains are constructed by random combination and sequential combination two ways (redundant event chains are filtered out). The training set, test set and validate set are divided as 8:1:1.

According to event chains in the training set, the event trigger evolutionary network is constructed, and there are 2,721 nodes and 21,298 edges. Table 1 shows statistics of the dataset.

Table 1. Statistic of the dataset.

News items	37,160
n	9
k	5
Event chains	3,281
Nodes	2,721
Edges	21,298

4.2 Baseline

We compare the proposed model with the following baseline models.

- Event Pair Based Model. Based on BiGRU, we can get the representations of history events in the event chain, $y_i, i = 1, 2, ..., n$, and the representations of candidate events, $y_{c_j}, j = 1, 2, \ldots, k$. Then, the weight of each history event is determined based on the following "Attention_1" mechanism between the candidate event and the history event, and further, we may get the score of the j_{th} candidate event s_j,

$$u_{ij} = ReLU(W_y y_i + W_c y_{c_j} + b_u) \quad (12)$$

$$\alpha_{ij} = \frac{exp(u_{ij})}{\sum_l exp(u_{lj})} \quad (13)$$

$$s_j = -||\sum_{i=1}^{n} \alpha_{ij} y_i - y_{c_j}||_2 \quad (14)$$

  where W_y, W_c, b_u are learnable parameters.
- Gated Graph Neural Network Model [10]. Based on the event representation y obtained by BiGRU, the event trigger evolutionary graph G, it uses GGNN to update the event representation. Similarly, in view of the updated event representations, Eq. 12–Eq. 14 are employed to calculate the score.

4.3 Experimental Setup

The lengths of more than 80% news titles are less than or equal to 11, so we set the title text length to 11. The learning rate is set to 0.0003, batch size is set to 64, and dropout is set to 0.3. The optimal model is remained by minimizing the loss of validation set. Except where noted, the weighted cross entropy loss function is used.

4.4 Discussion of Results

This paper uses Accuracy as evaluation metric. For a test sample, if the trained model selects the real next event from candidate events, where only a choice is correct, the sample is predicted correctly. As Eq. 15 shows, Accuracy is equal to the number of correctly predicted samples divided by the total number of test samples,

$$Accuracy = \frac{\#Samples\ which\ are\ predicted\ correctly}{\#All\ samples} \tag{15}$$

which can be regarded as *hits*@1 when the number of candidates is 5. Besides carrying out baseline models, we conduct ablation study to get a better insight into each component of the model. Experimental results are shown in Table 2.

Table 2. Comparison of results for different methods.

Method	Accuracy
Gated Graph Neural Network Model	48.62%
Event Pair Based Model+Attention_1	50.77%
Event Pair Based Model+Attention	50.77%
Event Pair Based Model+Attention+Trigger Representation	52.92%
Event Segment Based Model	54.77%
Event Segment Based Model+Trigger Representation	**55.38%**
Event Segment Based Model+Trigger Representation With General Update Way	52.62%
Event Segment Based Model+Trigger Representation-Weighted Loss	54.46%

"Gated Graph Neural Network Model" only models the transfer of events through the transfer of event triggers, which does not improve the performance. "Event Pair Based Model + Attention_1" corresponds to the first baseline. "Event Pair Based Model + Attention" is a special case of our model. We do not divide the event segments, regard the whole event chain as the same segment, and then use the "Attention" mechanism to calculate the weight of each history event. The experimental results show that there is no significant difference between the two attention mechanisms. "Event Pair Based Model + Attention + Trigger Representation" refers to performing "Attention" mechanism on the concatenation vector of the event semantic representation and the event trigger representation. Because the model learns the conversion patterns of actions from the event trigger evolutionary graph, it turns out that the performance is boosted. However, the above methods only calculate the score of the candidate event by integrating the correlation between each history event and the candidate event, without considering the connection of the same agent's successive actions in the history event chain.

"Event Segment Based Model" and "Event Segment Based Model + Trigger Representation" are the models we put forward. Their difference is whether to concatenate the event trigger representation after the event semantic representation or not. "Event Segment Based Model + Trigger Representation With General Update Way" denotes that in-degree neighbors are used to update the representations of triggers on the basis of the inherent model. "Event Segment Based Model + Trigger Representation-Weighted Loss" means that based on the inherent model, the weighted cross entropy loss function is changed into a general cross entropy loss function. "Event Segment Based Model + Trigger Representation" learns the semantics of event segments attach to different major subjects, considers the conversion relationship of triggers, in other words, makes full use of the agent and action information, thus has good performance. From the comparison results of ablation experiments, the event trigger representation learning method, event segmentation idea and sample weighting thought proposed in this paper are helpful for the prediction.

4.5 Case Study

This subsection illustrates prediction performance by giving some examples. The following event chains share the candidate event set. The first three event chains belong to the same topic, which is related to "THAAD"[6], and the last two event chains belong to the other topic, which is related to "national sovereignty".

History Event Chain 0: [[日本, 拟, 西南, 岛屿, 部署, 反导系统, 打破, 对华战略, 平衡], [美军, 派, 网络部队, 前往, 韩国, 防, 萨德反导系统, 泄密], [普京, 强硬, 反, 萨德, 警告, 美韩, 相应, 方式, 做出, 回应], [中方, 回应, 中俄, 是否, 北方四岛, 部署, 反导, 韩, 施压], [韩国, 外长, 称, 部署, 萨德, 韩美同盟, 决定, 韩方, 无意, 推翻], [韩媒, 否认, 中国, 曾, 多次, 侵略, 韩国, 势力, 刻意, 渲染], [韩, 称, 立即, 美, 协商, 部署, 剩余, 4, 架, 萨德, 发射车], [美国, 萨德, 连续, 两, 次, 测试, 成功, 韩, 总统, 下令, 追加], [中国, 急召, 韩, 大使, 反对, 追加, 萨德, 停止, 部署, 撤除, 已]]

History Event Chain 1: [[中国, 黄渤海, 军演, 一箭, 双雕, 震慑, 萨德, 入, 韩, 麻痹, 印度], [美, 要求, 韩国, 30日前, 完成, 部署, 萨德系统, 中方, 坚决反对], [俄媒, 称, 俄, 反导系统, 回应, 萨德, 中国, 走, 近], [韩国, 7日, 凌晨, 部署, 4, 辆, 萨德, 发射车, 中方, 表态], [韩国, 一面, 部署, 萨德, 一面, 求, 中国政府, 韩企, 减税], [美军, 4, 辆, 萨德, 发射车, 进入, 星州, 基地, 正式, 韩, 完成], [中方, 回应, 韩国, 完成, 部署, 萨德, 已, 韩方, 提出, 严正, 交涉], [韩国总统, 称, 部署, 萨德系统, 韩, 政府, 采取, 最佳, 措施], [韩媒, 称, 6, 辆, 萨德, 发射车, 全部, 完成, 部署, 已, 投入]]

History Event Chain 2: [[韩, 有意, 引进, 海上, 萨德, 导弹, 情报, 美日, 共享], [美国, 沙特, 部署, 萨德, 专家, 称是, 巩固, 军事优势], [中俄, 两军, 举行, 反导, 问题, 吹风会, 要求, 美韩, 撤出, 萨德], [驻韩美军, 座, 萨德系统, 炮台, 启用, 兵力, 体系, 部署, 完成], [韩国, 外长, 称, 考虑, 追加, 部署, 萨德反导系统], [韩国, 外长, 称, 考虑, 追加, 部署, 萨德, 重申, 萨德, 防御, 措施], [防空, 比武, 看, 中国, 反导, 能力, 正, 跻身, 世界, 一流], [中俄,

[6] https://en.jinzhao.wiki/wiki/Terminal_High_Altitude_Area_Defense.

两军, 12月, 举行, 空, 天安, 全, 反导, 计算机, 演习], [中国军方, 回应, 中俄, 反导, 演习, 共同, 应对, 地区, 导弹, 威胁]]

History Event Chain 3: [[安倍, 告知, 特朗普, 应对中国, 世纪, 最, 课题], [日本, 教材, 钓鱼岛, 篡改, 日本, 固有, 领土, 中方, 回应], [日本, 针对, 钓鱼岛, 祭出, 两, 狠招, 被指, 对内壮胆, 对外, 诈唬], [日本, 勾结, 美国, 谋占, 钓鱼岛, 再, 出, 花招, 中国, 需, 警惕], [俄防长, 俄军, 今年, 日俄争议岛屿, 部署, 师], [美日, 进行, 第12, 次, 联合, 夺岛, 训练, 目标, 竟, 针对, 钓鱼岛], [日本, 教科书, 篡改, 南京大屠杀, 历史, 中方, 回应, 称, 极其, 错误], [日本政府, 抗议, 韩国军队, 独岛, 周边, 海域, 开展, 军事训练], [日本, 教科书, 称, 南京大屠杀, 死亡人数, 确定, 中方, 驳斥]]

History Event Chain 4: [[外媒, 称, 安倍, 野心, 愈发露骨, 不遗余力, 加强, 钓鱼岛, 控制], [日本, 中学, 恢复, 拼刺刀, 课程, 引发, 反对, 声浪, 恐慌], [日本, 扩充, 海军, 故意, 挑起, 战端, 战舰, 平均, 吨位, 高于, 中国], [火上浇油, 韩军方, 重申, 朝鲜, 韩, 未, 收复, 地区], [韩国海军, 日韩争议岛屿, 周边, 训练, 日本, 提出, 抗议], [日本, 改, 教材, 妄称, 钓鱼岛, 领土, 中方, 警告, 其别, 挑衅], [印度, 梵文, 教科书, 竟, 妄称, 1962年, 战争, 印度, 击败, 中国], [印, 教科书, 造谣, 称, 印度, 1962年, 战胜, 中国, 遭, 美媒, 打脸], [日本, 学者, 公开, 古地图, 证实, 钓鱼岛, 中国领土, 图]]

Candidates: [[美称, 中国, 进行, 大规模, 军演, 试射, 20, 枚, 导弹, 模拟, 攻击], [部署, 6, 辆, 萨德, 发射车, 韩国, 引进, 海上, 萨德], [美国, 安全, 战略, 报告, 韩, 部署, 萨德, 辩护, 针对, 中国], [日, 教科书, 钓岛, 列为, 固有, 领土, 中方, 警告, 勿挑事], [外交部, 回应, 日, 史学家, 证明, 钓鱼岛, 中国, 固有, 领土]]

The following matrix shows the prediction scores of candidate events for each event chain. Rows denote event chains, and columns denote candidate events. The real labels for these event chains are 0, 1, 2, 3 and 4. In this case, history event chain 1, history event chain 2 and history event chain 3 are predicted correctly.

$$\begin{bmatrix} -9.4079 & -7.3515 & -7.1473 & -7.2442 & -8.1875 \\ -9.6748 & -5.9196 & -8.0906 & -8.0433 & -8.5127 \\ -9.2512 & -7.4936 & -6.8989 & 8.0224 & -9.1972 \\ -8.4414 & -8.1061 & -8.9752 & -7.0416 & -9.0865 \\ -10.5230 & -9.6750 & -9.4273 & -6.3349 & -7.3929 \end{bmatrix}$$

5 Conclusions

This paper introduces event trigger evolutionary graph and event segment to address event prediction problem on the Sina Military News corpus. Different from existing datasets for event prediction, there are not ready-made event chains for supervised learning in the corpus. Thus, from entity and keyword perspective, we carry out LDA topic model to identify the topics of news items, and then news items on the same topic in chronological order are segmented by a certain length window to obtain the short event chains. This paper defines event prediction as a multiple choice cloze task, and the candidate event set of each history event chain is prepared by sample combination. After then, the trainable data are got. We construct event trigger evolutionary graph, and learn trigger aspect transition patterns among events for helping predicting the next possible action. We learn features of event segments divided by event subjects, and calculate relations between the candidate event and these event segments so as to get the score of the candidate event. The comparison results demonstrate the effectiveness of the proposed model. There is still a long way to go for event prediction research. In the future, we try to improve prediction accuracy and apply our method to news corpora in other domains. Another important direction is to deal with event generation problem.

Acknowledgements. This research was supported by National Natural Science Foundation of China under Grant Nos. 71731002 and 71971190. The computations were partly done on the high performance computers of State Key Laboratory of Scientific and Engineering Computing, Chinese Academy of Sciences.

References

1. Aggarwal, K., Theocharous, G., Rao, A.B.: Dynamic clustering with discrete time event prediction. In: Proceedings of the 43rd International ACM SIGIR Conference on Research and Development in Information Retrieval, pp. 1501–1504 (2020)
2. Allan, J., Papka, R., Lavrenko, V.: On-line new event detection and tracking. In: Proceedings of the 21st Annual International ACM SIGIR Conference on Research and Development in Information Retrieval, pp. 37–45 (1998)
3. Ammanabrolu, P., Cheung, W., Broniec, W., Riedl, M.O.: Automated storytelling via causal, commonsense plot ordering. In: Proceedings of the AAAI Conference on Artificial Intelligence, vol. 35, pp. 5859–5867 (2021)
4. Granroth-Wilding, M., Clark, S.: What happens next? Event prediction using a compositional neural network model. In: Proceedings of the Thirtieth AAAI Conference on Artificial Intelligence, pp. 2727–2733 (2016)
5. Heinrich, K., Zschech, P., Janiesch, C., Bonin, M.: Process data properties matter: introducing gated convolutional neural networks (GCNN) and key-value-predict attention networks (KVP) for next event prediction with deep learning. Decis. Support Syst. **143**, 113494 (2021)
6. Hu, L., Li, J., Nie, L., Li, X.L., Shao, C.: What happens next? Future subevent prediction using contextual hierarchical LSTM. In: Proceedings of the Thirty-First AAAI Conference on Artificial Intelligence, pp. 3450–3456 (2017)
7. Hu, L., Yu, S., Wu, B., Shao, C., Li, X.: A neural model for joint event detection and prediction. Neurocomputing **407**, 376–384 (2020)
8. Lei, L., Ren, X., Franciscus, N., Wang, J., Stantic, B.: Event prediction based on causality reasoning. In: Nguyen, N.T., Gaol, F.L., Hong, T.-P., Trawiński, B. (eds.) ACIIDS 2019. LNCS (LNAI), vol. 11431, pp. 165–176. Springer, Cham (2019). https://doi.org/10.1007/978-3-030-14799-0_14
9. Li, M., et al.: Future is not one-dimensional: graph modeling based complex event schema induction for event prediction. arXiv preprint arXiv:2104.06344 (2021)
10. Li, Z., Ding, X., Liu, T.: Constructing narrative event evolutionary graph for script event prediction. In: Proceedings of the 27th International Joint Conference on Artificial Intelligence, pp. 4201–4207 (2018)
11. Liu, X., Huang, H.Y., Zhang, Y.: Open domain event extraction using neural latent variable models. In: Proceedings of the 57th Annual Meeting of the Association for Computational Linguistics, pp. 2860–2871 (2019)
12. Luo, W., et al.: Dynamic heterogeneous graph neural network for real-time event prediction. In: Proceedings of the 26th ACM SIGKDD International Conference on Knowledge Discovery & Data Mining, pp. 3213–3223 (2020)
13. Lv, J., et al.: HGEED: hierarchical graph enhanced event detection. Neurocomputing **453**, 141–150 (2021)
14. Lv, S., Qian, W., Huang, L., Han, J., Hu, S.: SAM-Net: integrating event-level and chain-level attentions to predict what happens next. In: Proceedings of the AAAI Conference on Artificial Intelligence, vol. 33, pp. 6802–6809 (2019)
15. Mao, Q., et al.: Event prediction based on evolutionary event ontology knowledge. Futur. Gener. Comput. Syst. **115**, 76–89 (2021)

16. Ning, Q., Subramanian, S., Roth, D.: An improved neural baseline for temporal relation extraction. In: Proceedings of the 2019 Conference on Empirical Methods in Natural Language Processing and the 9th International Joint Conference on Natural Language Processing (EMNLP-IJCNLP), pp. 6203–6209 (2019)
17. Peng, H., et al.: Fine-grained event categorization with heterogeneous graph convolutional networks. In: Proceedings of the Twenty-Eighth International Joint Conference on Artificial Intelligence, pp. 3238–3245 (2019)
18. Su, Z., Jiang, J.: Hierarchical gated recurrent unit with semantic attention for event prediction. Futur. Internet **12**(2), 39 (2020)
19. Sun, W., Wang, Y., Gao, Y., Li, Z., Sang, J., Yu, J.: Comprehensive event storyline generation from microblogs. In: Proceedings of the ACM Multimedia Asia, pp. 1–7 (2019)
20. Taymouri, F., Rosa, M.L., Erfani, S., Bozorgi, Z.D., Verenich, I.: Predictive business process monitoring via generative adversarial nets: the case of next event prediction. In: Fahland, D., Ghidini, C., Becker, J., Dumas, M. (eds.) BPM 2020. LNCS, vol. 12168, pp. 237–256. Springer, Cham (2020). https://doi.org/10.1007/978-3-030-58666-9_14
21. Wang, R., Zhou, D., He, Y.: Open event extraction from online text using a generative adversarial network. In: Proceedings of the 2019 Conference on Empirical Methods in Natural Language Processing and the 9th International Joint Conference on Natural Language Processing (EMNLP-IJCNLP), pp. 282–291 (2019)
22. Wang, Z., Zhang, Y., Chang, C.Y.: Integrating order information and event relation for script event prediction. In: Proceedings of the 2017 Conference on Empirical Methods in Natural Language Processing, pp. 57–67 (2017)
23. Xu, J., Wang, H., Niu, Z.Y., Wu, H., Che, W., Liu, T.: Conversational graph grounded policy learning for open-domain conversation generation. In: Proceedings of the 58th Annual Meeting of the Association for Computational Linguistics, pp. 1835–1845 (2020)
24. Yang, S., Feng, D., Qiao, L., Kan, Z., Li, D.: Exploring pre-trained language models for event extraction and generation. In: Proceedings of the 57th Annual Meeting of the Association for Computational Linguistics, pp. 5284–5294 (2019)
25. Yang, Y., Wei, Z., Chen, Q., Wu, L.: Using external knowledge for financial event prediction based on graph neural networks. In: Proceedings of the 28th ACM International Conference on Information and Knowledge Management, pp. 2161–2164 (2019)

Aspect Based Fine-Grained Sentiment Analysis for Public Policy Opinion Mining

Yueming Zhao[1], Ying Li[1], Yijun Liu[2,3], and Qianqian Li[2,3](✉)

[1] College of Computer Science and Technology, Key Laboratory of Symbol Computation and Knowledge Engineering of Ministry of Education, Jilin University, Changchun 130012, China
[2] Institutes of Science and Development, Chinese Academy of Sciences, Beijing 100190, China
[3] School of Public Policy and Management, University of Chinese Academy of Sciences, Beijing 100049, China
lqqcindy@casipm.ac.cn

Abstract. Fine-grained mining has substantial practical significance for understanding public opinion toward public policies and optimizing relevant decision-making processes. In this study, an aspect-based sentiment analysis model integrated with both a sentiment lexicon for the policy domain and mutual information (MI) is established from the perspective of attitudinal orientations toward several elements of public policies, including subjects, objects, and tools to analyze comments. First, terms related to the aspects of comments on different policy elements are extracted using the association rules in conjunction with the word2vec algorithm. In addition, the contextual neighbor principle is employed to extract terms that express evaluative opinions and to extricate aspect-based two-tuples from comments. Subsequently, the sentiment lexicon for the policy domain is expanded using the sentiment orientation pointwise mutual information (SO-PMI) algorithm. Four machine learning models (i.e., the support vector machine, logistic regression, naive Bayes, and k-nearest neighbors models) and two deep learning models (i.e., the convolutional neural network (CNN) and long short-term memory models) are compared in terms of their classification performance based on 87,304 comments on public policies crawled from the Weibo platform. The results show that the CNN model integrated with the domain-specific sentiment lexicon and mutual information is effective at facilitating policy element-oriented fine-grained sentiment analysis of public opinion toward policies and therefore has practical decision-making value for the government to better understand the policy demands of the public.

Keywords: Public policy · Sentiment analysis · Fine-grained · CNN · SO-PMI

1 Introduction

Public policies are instruments that authoritatively apportion social benefits and therefore have a direct impact on the interests of the public. Public participation and support collectively comprise the "public nature" of public policies [1]. Political research has shown that there exists a substantial positive correlation between public opinion and policies and

J. Chen et al. (Eds.): KSS 2022, CCIS 1592, pp. 202–217, 2022.
https://doi.org/10.1007/978-981-19-3610-4_15

that public opinion can influence policy formulation [2]. Advances in information and communication technologies in recent years have allowed for continual improvements to relevant platforms (e.g., petition websites, social media, and open government portals) and led to the creation of new avenues for and cost reduction of public participation in government decision-making processes. The emergence of extensive public opinion data provides a reference for the government to make scientifically sound decisions [3]. Moreover, the government can better understand public expectations of policies through public opinion data [4]. Between 2005 and 2017, the U.S. National Science Foundation and Cornell University jointly conducted the multidisciplinary collaborative Cornell e-Rulemaking Initiative through the establishment of an online participation platform for policies, termed RegulationRoom (www.RegulationRoom.org). This platform was used to publish information regarding proposed government regulations on which the public commented. Decision-making agencies subsequently optimized their decisions by analyzing these comments.

The growing number of public opinion-related risk incidents precipitated by public policies in recent years has prompted China to pay increasing attention to the decision-making value of public opinion on policies. On September 1, 2019, the State Council of China promulgated Provisional Regulations on Major Administrative Decision-making Procedures. The regulations state that we must insist on scientific, democratic and law-based decision-making, strive to improve the quality and efficiency of major administrative decisions, with a clear mandate to adequately exploit the role of "public opinion tracking". In 2020, the Central Committee of the Chinese Communist Party issued an Implementation Outline for the Establishment of a Society Governed by the Rule of Law (2020–2025). In this outline, "strengthening the mechanism that allows the public to participate in major public decision-making processes" was listed as the first article related to the protection and strengthening of the rights of citizens. However, big data on public opinion on the Internet, known for their mixed content, structural heterogeneity, and semantic sparseness, which poses a huge challenge to policy makers in value mining to understand public needs and policy preferences.

Topic analysis, statistical analysis of word frequencies and sentiment analysis are the most common used methods currently to extract the decision-making value of policy opinion [5, 6]. These methods primarily classify the opinions of users and analyze their sentiment polarity at a document level. Practically, government administrators are more interested in fine-grained sentiment analysis. Here the fine-grained sentiment analysis refers to detect sentiment polarity in different granularities of aspects. Aspect-based opinion mining is a fine-grained level of opinion mining, revealing the sentiment polarity towards the policy aspects and showing the explanations why citizens pro-/con- the released policy. The development of sentiment analysis techniques has enabled policy makers to determine public sentiment towards policy 'elements' through 'fine-grained' sentiment analysis methods. Aspect-based sentiment analysis begins by extracting the entity and the text that evaluates it, and then analyses people's emotional disposition towards the entity. This analytical process involves the extraction of "aspect(The entity being evaluated)" and "evaluation(words to evaluate the entity)",then create the <aspect, evaluation> binary. As shown in Fig. 1, following the above approach, two combinations can be extracted: <government, making efforts> and <illegal enterprises, shut

down>. This shows the difference in public attitudes towards the government and illegal enterprises.

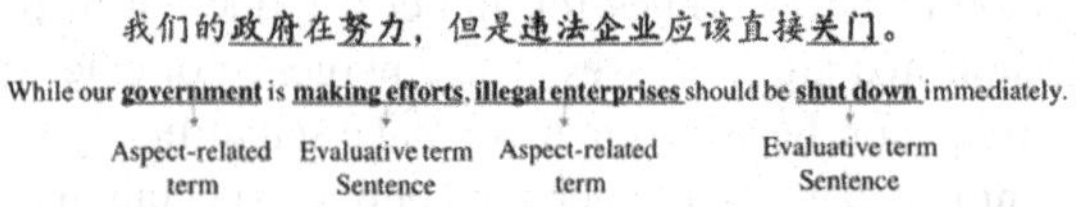

我们的政府在努力，但是违法企业应该直接关门。

While our government is making efforts, illegal enterprises should be shut down immediately.

Aspect-related term | Evaluative term Sentence | Aspect-related term | Evaluative term Sentence

Fig. 1. Example of extraction of "aspect-based" evaluative collocations from public opinion toward public policies

According to public policy theories, a policy consists of three principal elements: its subjects, objects, and implementation tools. The subjects of a policy refer primarily to the individuals, groups, or organizations that participate in its policy. The objects of a public policy refer to the targets on which it acts and its scope of influence, that is, the social problems that the policy is designed to tackle or the target population. The tools of a policy refer to the mechanisms, means, and methods that the government employs to achieve specific policy goals. Hence, the elements of a policy are categorized into four types: policy-makers, policy executers/implementers, those affected by the policy, and policy implementation tools. Then, based on the above policy elements, this study performs a "fine-grained" aspect-based sentiment analysis of public opinion toward public policies. The contributions of this study can be summarized as following: First, we establish a policy opinion domain lexicon based on SO-PMI algorithm. Second, we distinguish the policy opinion aspects according to the policy elements. Third, we compare the machine learning and deep learning models for the policy aspect-based opinion analysis. The results show that the aspect-based sentiment classification methods integrated with the sentiment lexicon for the policy domain and mutual information display better classification performance. The "fine-grained" sentiment analysis model for public opinion on policies introduced in this study can provide a more accurate analysis of public opinion. It supports decision-makers in understanding the key areas of policies that require adjustments.

2 Relevant Literature

2.1 Social Media and Government Policies

In recent years, the number of studies exploring the relationship between social media and government decision-making has steadily increased as public participation in public policy decisions has been increasingly influenced by social media. At the beginning of the 21st century, social media was considered an effective tool to promote two-way communication and improve the level and quality of public participation in politics through the participation of a relatively small number of users [7]. Compared to conventional opinion solicitation mechanisms, social media allows the government to obtain feedback from more individuals. Sobkowicz et al. [8] studied the process by which social media influences the dynamics of thought of the public and modeled an opinion network based on content analysis of social media. This model helps policy-makers assess the impact of specific policies on people throughout society and improve their impact. Based on the latent Dirichlet allocation (LDA) topic model, Driss et al. put forward a semantic analysis framework that integrates the mining of the decision-making value of public opinion in social media with the policy agenda setting and policy evaluation stage to analyze the policy needs of the public [6]. These studies all suggest that an understanding of the dynamics of public thought is important for formulating and improving policies.

2.2 Fine-Grained Sentiment Analysis

There are three levels of accuracy for text sentiment mining: (from low to high) the document, sentence, and aspect levels. Document-level sentiment mining classifies the sentiments expressed in an entire document into three types, namely, positive, neutral, and negative [9]. This classification is too broad and precludes an adequate understanding of the connotative sentiments expressed in the document. Sentence-level sentiment mining treats sentences as units for sentiment expression and calculates the sentiment orientations they express. However, the sentiment expressed in a sentence is unable to reflect the sentiment attribute of the objects of evaluation. Let us take the sentence in Fig. 1 as an example: "While our government is making efforts, illegal enterprises should be shut down immediately." The commentator has a positive attitude toward the government but makes a negative comment on the other object of evaluation, that is, illegal enterprises. Fine-grained sentiment analysis cannot be achieved at the document and sentence levels. The first step in aspect-level sentiment analysis extracts evaluative words and the entities being evaluated. The second step classifies these evaluative words. The third step is to analyse the emotional tendencies of the whole sentence [10]. This method is able to extract "fine-grained" decision-making value from public opinion.

Aspect-based sentiment analysis involves the extraction of aspect-related and evaluative terms. In terms of the extraction of aspect-related terms, Jakob and Gurevych transformed this process into a sequence labeling problem using conditional random fields (CRFs) and found that the CRF-based approach outperformed a supervised learning approach (used to yield baselines) on datasets from the network service and electronic product domains [11]. By introducing features such as part-of-speech (POS) based on CRFs, Zhang et al. developed a training model, which, when paired with a domain-specific lexicon, yielded relatively good extraction results [12]. Xu et al. trained models by introducing more shallow grammatical features (e.g., contextual information and positions) in addition to POS [16]. Alternatively, aspects can also be extracted using topic models such as the LDA [13] and probabilistic latent semantic analysis [14] models. This method extracts corresponding aspects based on the relationship between the document and words. By exploiting the fact that aspect-related terms are mostly nouns and noun phrases, Hu and Liu [15] produced POS tags for product review data and used the Apriori association rule mining algorithm to identify the frequent nouns and noun phrase items as candidate aspects, which were subsequently pruned to form an aspect database.

Evaluative terms can be extracted based on rules, statistical models, or deep learning models. Rule-based extraction methods primarily mine the content of opinions using lexicons. WordNet and HowNet are among the commonly used opinion evaluation lexicons. The identification accuracy for evaluative terms can be improved through means such as the calculation of semantic similarity [16] and the use of PMI [17]. Extraction of opinions based on statistical models is achieved through modelling based on the relationship between aspect-related and evaluative terms. For example, Lin et al. developed a joint sentiment–topic (JST) model and a reverse JST model by improving the LDA model using the Bayesian model [18]. The hidden Markov model with LDA introduced by Duric and Song is another example [19]. Extraction based on deep learning models involves obtaining latent semantic features through learning from massive volumes of data. For example, Huang et al. performed POS tagging and identified named entities using a bidirectional LSTM network combined with a CRF layer [20].

3 Data Collection and Preprocessing

A total of 87,304 comments on major public policies issued in recent years (as shown in Table 1), including those that sparked heated public debate (e.g., the raising of the threshold for individual income tax, the introduction of a cooling-off period for divorce, and the Vaccine Management Law), were collected from Sina Weibo using the crawling technique.

Table 1. Basic information on the public policies examined in this study

No	Public policy	Date of issuance	No	Public policy	Time of issuance
1	Raising the threshold for individual income tax	2018–06	5	Compulsory waste sorting	2019–01
2	"Menstruation leave" planned to be introduced in Shandong	2018–07	6	Blacklisting of people who spread rumors online as dishonest subjects	2019–07
3	Introduction of a cooling-off period for divorce	2018–08	7	Prohibition of tour companies from big data-enabled price discrimination against existing customers	2019–10
4	Vaccine Management Law	2018–11	8	Improvement of laws for minors	2019–11

In this study, the data preprocessing procedure mainly involves the cleansing of comment data (e.g., word segmentation and elimination of stop words). The list of stop words (containing 1,895 words) used in this process was created through the integration of the stop-word lexicons produced by Sichuan University and Harbin Institute of Technology and the Baidu list of stop words.

4 Aspect-Based Sentiment Analysis Algorithm for Public Opinion on Public Policies

4.1 Extraction of Aspect-Related Terms

4.1.1 Policy Element-Oriented Classification of Aspect-Related Terms

Extraction of aspect-related terms is the first step of aspect-based sentiment analysis. Aspect-related terms are the smallest objects of a certain entity of evaluation and consist primarily of nouns or noun phrases [15]. As mentioned in the Introduction, the evaluation involved in public opinion on public policies is categorized into four aspects: policy-makers, policy executors, those affected by the policy, and policy tools. Table 2 summarizes the classified aspects of comments on public policies and gives examples of aspect-related terms.

Table 2. Classification of aspect-related terms in the public policy domain

Aspect	Example
Policy-makers	Governments and countries
Policy executors/implementers	Enterprises and sectors
Those affected by the policy	General public and specific groups of people
Policy tools	Fines and reviews

4.1.2 Extraction of Basic Aspect-Related Terms Based on the Apriori Algorithm

The Apriori association rule algorithm [15] is used in this study to extract keywords related to policy elements. Nouns and noun phrases are used to form a candidate lexicon. Item sets supported by more than 1% of the comments are obtained from text-based comment data. In other words, each item set that appears at least once in every hundred sentences is treated as a frequent itemset. All the frequent itemsets yielded through the iteration of the algorithm are treated as candidate aspect-related terms.

4.1.3 Word2vec-based Expansion of Terms Related to Aspects of the Policy Domain

After the basic aspect-related term lexicon is constructed, word2vec word embedding is used to represent the terms involved in public opinion toward policies. Extended terms semantically similar to the terms in the basic lexicon are identified through the mapping of the terms in the basic lexicon to a higher-dimensional space and the subsequent calculation of their correlations based on their cosine similarity. Cosine similarity is a measure of the similarity between two vectors based on their cosine in the vector space. A cosine close to 1 indicates that the two vectors are similar to each other. Conversely, a cosine close to 0 suggests that the two vectors are dissimilar. The cosine of two vectors is calculated using Eq. (1).

$$\cos(\theta) = \frac{a \cdot b}{||a| \times |b||} = \frac{\sum_{i=1}^{n} (x_i \times y_i)}{\sqrt{\sum_{i=1}^{n} (x_i)^2} \times \sqrt{\sum_{i=1}^{n} (y_i)^2}} \tag{1}$$

where a and b are the n-dimensional vectors of the first and second terms, respectively. $\vec{a} = (x_1, x_2, \cdots, x_n), \vec{b} = (y_1, y_2, \cdots, y_n)$, where x_i, y_j are float number indicating the frequency of each term inside a document, while the dimension of the vector corresponds to the terms available in the document. In this study, the similarity between the candidate terms and the basic aspect-related terms is calculated.

4.2 Extraction of Evaluative Terms

After the terms being evaluated are determined, it is necessary to extract the related evaluative words. Based on the POS tags of terms, adjectives, adverbs, idioms, and expressions (i.e., habitually used phrases in human languages) are selected as candidate evaluative terms, from which the term that is the closest to the aspect-related term of interest is identified based on position indexes. Specifically, an aspect-related term is first located, and then its distance from each key candidate evaluative term is determined. We choose the evaluative word closest to the aspect-related word as the word to evaluate the entity.On this basis, <aspect-related term, evaluative term> evaluative collocations are formed. Thus, the extraction of evaluative collocations is completed. Table 3 gives some examples of evaluative collocations. In this study, 600 negative collocations and 581 positive collocations are extracted.

Table 3. Examples of extracted evaluative collocations in the public policy domain

Emotional orientation	Evaluative collocation
Positive	<policy, effective>, <country, trustworthy>, <government, responsible>, <official, strict>, <policy, heartwarming>
Negative	<fine, little>, <the people, pitiful>, <official, irresponsible>, <government, lazy>, <policy, useless>

4.3 SO-PMI-Based Construction of a Lexicon Specific to Public Opinion on Policies

All the available sentiment lexicons, such as the CNKI HowNet Sentiment Lexicon and the Chinese sentiment vocabulary ontology, provide only basic sentiment terms and fail to cover all the sentiment terms involved in the domain of comments on public policies. For example, terms, such as “走过场” (goes through the formality) and “形式主义”

(formalism) are not included in the basic sentiment lexicons. In this study, SO-PMI is used to extend the lexicon oriented to the domain of public opinion on public policies. This algorithm determines the sentiment orientation of a term based on PMI, specifically its closeness to the positive and negative seed words. If the SO-PMI for a term exceeds the threshold, then the term is closer to the positive seed term and more likely to be a positive term and is therefore determined to be a positive sentiment term. If the SO-PMI for a term is below the threshold, then the term is more likely to be a negative word and is therefore determined to be a negative sentiment term.

PMI refers to the probability that two terms appear in the same piece of text. A high probability means that the two terms are likely to have the same sentiment orientation. PMI is calculated using the following equation:

$$PMI(w_1, w_2) = \log_2 \frac{P(w_1,w_2)}{P(w_1)P(w_2)} \tag{2}$$

where $P(w_1,w_2)$ is the probability that two terms, w_1 and w_2, appear concurrently and $P(w_1)$ is the probability that w_1 appears. As most Weibo texts are short in length, it is highly likely that the seed and candidate sentiment terms do not appear at the same time; that is, the mutual information is zero, which has a negative impact on the result. To address this issue, the Laplace smoothing technique is used to improve the PMI algorithm. Specifically, the equation used to calculate $P(w_1, w_2)$ in Eq. (2) is altered to the form shown in Eq. (3).

$$P(w_1, w_2) = \frac{count(w_1,w_2)+1}{n+2} \tag{3}$$

where $count(w_1, w_2)$ is the number of times that w_1 and w_2 appear in the same comment and n is the total number of comments.

SO-PMI is calculated with a set of positive terms and a set of negative terms as reference terms. The sentiment orientation of a term w can be determined based on the difference between the PMI between w and the positive terms and the PMI between w and the negative terms.

SO-PMI is calculated using the following equation:

$$\mathrm{SO-PMI}(w) = \sum_{Pwd \in Pwords} PMI(w, Pwd) - \sum_{Nwd \in Nwords} PMI(w, Nwd) \tag{4}$$

where *Pwords* is the set of positive seed terms, *Pwd* is each positive term in *Pwords*, *Nwords* is the set of negative seed terms, and *Nwd* is each negative term in *Nwords*. A term with a positive SO-PMI is positively oriented, whereas a term with a negative SO-PMI is negatively oriented.

A sentiment lexicon for comments in the public policy domain is constructed using the following procedure. First, terms functioning as adjectives, including verbs, nouns, adverbs, and other expressions (i.e., habitually used terms in human languages), are extracted based on the POS tags of the terms segmented from comments and subsequently treated as candidate sentiment terms. Second, sentiment terms with an intensity greater than 5 (these terms have a more marked sentiment orientation) in the Chinese sentiment vocabulary ontology and the CNKI HowNet Lexicon are combined. After the duplicates are removed, 5,381 basic positive sentiment terms and 11,337 basic negative sentiment terms are obtained. Third, the candidate sentiment terms and the terms in the basic sentiment lexicon are compared to remove duplicates. Subsequently, the term frequency-inverse document frequency (TF-IDF) score of each term is calculated. Specifically, the TF-IDF score of term i in document j, $tf - idf_{i,j}$, is calculated as follows: $tf - idf_{i,j} = tf_{i,j} \times idf_i$, where

$$tf_{i,j} = \frac{n_{i,j}}{\sum_k n_{k,j}} \tag{5}$$

where $n_{i,j}$ is the number of times term i appears in document j and $\sum_k n_{k,j}$ is the number of times that all the terms in document j appear.

$$idf_i = log\frac{|D|}{\left|\{j : t_i \in d_j\}\right| + 1} \tag{6}$$

where $|D|$ is the total number of documents in the corpus and $\left|\{j : t_i \in d_j\}\right|$ is the number of documents that contain the term in question. Analysis of the equation used to calculate $tf - idf_{i,j}$ reveals that a term with a high TF-IDF score is characterized by a high frequency of occurrence in specific documents and a low frequency of occurrence in the whole corpus.

Afterwards, the sentiment terms are sorted in descending order of their TF-IDF scores. Then, 15 positive seed terms and 15 negative seed terms are selected from the top 100 terms. Finally, a total of 138 negative sentiment terms and 112 positive sentiment terms are obtained using the improved SO-PMI algorithm. Table 4 gives examples of the terms in the basic sentiment lexicon and the sentiment lexicon specific to the public policy domain.

Table 4. Examples of extracted terms in the sentiment lexicon

Lexicon	Sentiment orientation	Example
Basic sentiment lexicon	Positive	Good, kind, happy, effective, very important, perfect
	Negative	Inferior, superfluous, immature, heart-rending, shameless, annoying
Sentiment lexicon specific to the public policy domain	Positive	Expectation, make an effort, good deed, useful, benefit, proud, assurance

(*continued*)

Table 4. (*continued*)

Lexicon	Sentiment orientation	Example
	Negative	Perfunctory, treat the symptoms but not the root cause, tokenism, laws are not observed, talk in a bureaucratic tone, one size fits all, empty talk, useless

4.4 MI-Enhanced Calculation of Classification Probabilities

MI is a measure of the mutual dependence between two random variables and indicates whether there exists a relationship between them and, if so, its strength. Equation (7) defines the degree of association (i.e., MI) between a certain feature term m and a certain category s.

$$MI(m, s) = \log_2 \frac{p(m, s)}{p(m) * p(s)} \approx \log_2 \frac{N(m, s) * T}{N(m) * N(s)} \tag{7}$$

where $N(m, s)$ is the number of times when feature term m appears in category s, T is the total number of documents in the database, $N(m)$ is the number of times feature term m appears in the database, and $N(s)$ is the number of documents belonging to category s in the document.

In an evaluation collocation, there may be more than one evaluative term describing the same entity, so the joint effect of these multiple evaluative terms is taken into account in the calculation. Equation (8) calculates the total mutual information of all evaluation terms in an evaluation collocation.

$$MI(c, s_i) = \sum_{j=1}^{n} \max\{MI\left(p_j, s_i\right), 0\}, \; s_i \epsilon \{0, 1\} \tag{8}$$

where c is an evaluative term and p_j is the probability that the evaluative collocation two-tuples of which c is a constituent belongs to category i. Category set s = $\{0, 1\}$, where 0 and 1 represent the negative and positive categories, respectively.

The MI integration approach allows for the calculation of the MI between the aspect-related term in each evaluative collocation and the positive and negative categories. MI is set to 0 when it is negative. In addition, it is necessary to calculate the set of probabilities that the evaluative collocation of interest belongs to the negative or positive category, $\{p_1, p_2\}$, using the machine learning method. Then, the total MI for the evaluative collocation is calculated using Eq. (8) and is subsequently used, together with Eq. (9), to calculate the weighted classification probability, based on which the classification result is determined.

$$p(s_i|c) = p_i \frac{MI(c, s_i)}{\sum_{i=0}^{1} MI(c, s_i)} \tag{9}$$

where p_i is the probability that the evaluative collocation belongs to the negative or positive category yielded by the reference classifier that does not include MI. If $p(s_0|c) =$

$p(s_1|c) = 0$, $p(s_0|c) = p(s_1|c)$ is set to 1, and the classification result yielded by the reference classifier prevails.

Table 5 compares the classification probabilities produced by the KNN classifier (used as a reference classifier) and its variant integrated with MI. Here, the <country, difficult> evaluative collocation in Table 5 is used as an example. According to all the sentiment lexicons (e.g., the CNKI HowNet Sentiment Lexicon), the term "difficult" has a negative connotation. However, when paired with the subject "country" to form an evaluative collocation, the term "difficult" denotes that there are obstacles to the governance of the country and expresses a positive sentiment from the perspective of risks associated with public opinion. The addition of MI allows for an adjustment to the probabilities of negative and positive sentiment orientations yielded by the conventional classifier based on the difference between evaluative collocations. Evidently, the integration of MI is effective at rectifying some erroneous judgments given by the conventional classifier and therefore improves its predictive accuracy.

Table 5. Classification probabilities produced by traditional classifier and MI-based method

Sentiment orientation	Evaluative collocation	Reference classifier (KNN)		MI-based KNN classifier	
		Negative	Positive	Negative	Positive
Positive	<Country, difficult>	0.67	0.33	0.20	**0.24**
	<Relevant authority, strict>	0.67	0.33	0.20	**0.24**
Negative	<Fine, light>	0.33	0.67	**0.32**	0.05
	<Regulator, make a fortune>	0.33	0.67	**0.31**	0.03

5 Experimental Results

5.1 Model Evaluation Metrics

In this study, precision, recall, and F1-score are used to evaluate model performance. These three metrics are formally described below:

$$Precision = \frac{TP}{TP + EP} \quad (10)$$

$$Recall = \frac{TP}{TP + FN} \quad (11)$$

$$F1 = \frac{2 \times TP}{2 \times TP + FP + FN} \quad (12)$$

5.2 Baseline Methods

5.2.1 Machine Learning Classification Models

In this study, four commonly used machine learning L classification models, namely, the SVM, NB, LR, and KNN models, are selected for analysis. To determine their parameters, the word2vec algorithm is first used to generate 100-, 150-, and 200-dimensional vectors to compare the effects of term vector dimensionality on their classification performance. Based on the difference in the classification performance in Table 6, 150-dimensional vector representations are used in this study. Moreover, the K value in the KNN model also affects the classification results. A comparison of the classification and evaluation results obtained at K values of 3, 4, and 5 (Table 7) reveals that the best results are achieved at a K value of 3. Hence, the term vector dimensionality and the K value in the KNN classifier are set to 150 and 3, respectively.

Table 6. Effects of vector dimensionality on the SVM algorithm

Vector dimensionality	Precision	Recall	F1 score
100	84.34%	83.33%	83.35%
150	**86.20%**	**85.88%**	**85.76%**
200	83.90%	82.77%	82.78%

Table 7. Effects of K value on the KNN algorithm

K value	Macroaverage	Microaverage
3	**86.95%**	**85.31%**
4	85.79%	83.90%
5	82.94%	82.49%

5.2.2 Deep Learning Classification Models

In recent years, deep learning has gradually been applied to sentiment analysis research. In this method, a neural network is established to simulate the way that the human brain thinks to analyze and study data such as texts and images. The CNN and LSTM models are among the commonly used deep learning models and have recently shown exceptional performance in text processing [21]. Hence, these two models are used in this study as reference deep learning classification models.

5.3 Experimental Results

Table 8 summarizes the evaluation results for the classification performance of the models for the negative and positive aspects of the comments on public policies. It is evident that integration of the policy domain-specific lexicon and MI considerably improves the classification performance of the models. Here, the classification results for the negative aspect are discussed as an example. Of all the models, the LR model benefits the most from the integration of the lexicon and MI in terms of its classification performance. Specifically, this addition improves precision, recall, and F1-score for the LR model by 10.17%, 20.72%, and 17.46%, respectively. The reference machine learning classifiers generally perform relatively poorly in terms of recall. However, integration of the lexicon and MI significantly improves the recall for the SVM, LR, NB, and KNN machine learning classifiers and increases the performance metrics for the CNN and LSTM classifier by 2–8%.

Table 8. Classification performance of the models for the negative and positive aspects

Classification model	Performance evaluation for the negative aspect			Performance evaluation for the positive aspect		
	Precision	Recall	F1-score	Precision	Recall	F1-score
SVM	88.60%	60.62%	72.00%	63.84%	88.80%	74.29%
LR	86.60%	57.00%	68.75%	62.01%	88.20%	72.82%
NB	72.30%	67.88%	70.05%	64.16%	68.90%	66.47%
KNN (K = 3)	72.80%	69.43%	71.09%	73.68%	60.80%	66.67%
CNN	82.00%	84.00%	83.00%	81.48%	79.20%	80.37%
LSTM	85.50%	78.51%	81.90%	79.20%	86.00%	82.50%
SVM (lexicon + MI)	**96.80%**	79.79%	87.50%	79.19%	96.80%	87.15%
LR (lexicon + MI)	96.70%	77.72%	86.21%	77.72%	**97.50%**	86.50%
NB (lexicon + MI)	90.40%	83.42%	86.79%	81.82%	89.40%	85.46%
KNN (lexicon + MI)	87.70%	81.87%	84.72%	86.99%	78.80%	82.74%
CNN (lexicon + MI)	90.40%	86.70%	**88.50%**	**88.20%**	92.60%	**90.30%**
LSTM (lexicon + MI)	89.20%	**86.80%**	86.80%	85.40%	87.30%	87.70%

The training process of LSTM neural network is shown in Fig. 2. LSTM neural network has been iterated for 50,000 times. Figure 2 (a) shows the loss rate change curve every 50 iterations, and Fig. 2 (b) shows the accuracy change curve every 50 iterations. The blue line in the figure indicates the actual data, and the orange line indicates the change curve formed after fitting the real data. It can be seen in Fig. 2 that the loss rate of the deep learning classifier gradually decreases and the accuracy rate gradually increases during the training process, and finally the accuracy rate can reach 93.75%, which shows that the training model has a good fitting degree.

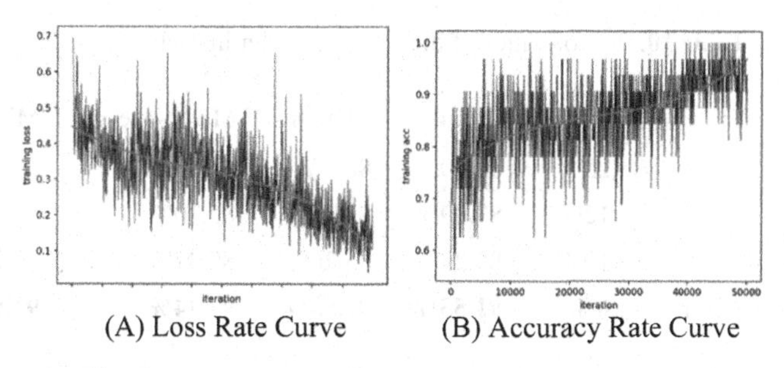

(A) Loss Rate Curve (B) Accuracy Rate Curve

Fig. 2. LSTM training set loss rate and accuracy curve

Analysis of the overall performance evaluation in Table 9 reveals that the CNN algorithm integrated with MI and the sentiment lexicon performs best, with a precision of nearly 90%. Overall, the deep learning classifiers perform considerably better than the conventional machine learning classifiers, with an increase of approximately 10% in both their macro- and microaverages after integration.

Table 9. Overall performance evaluation

Classifier	Macroaverage	Microaverage
SVM	78.79%	74.86%
LR	75.03%	71.75%
NB	68.27%	68.36%
KNN (K = 3)	72.23%	72.32%
CNN	81.76%	81.78%
LSTM	82.39%	82.20%
SVM (lexicon + MI)	88.00%	87.01%
LR (lexicon + MI)	87.58%	86.44%
NB (lexicon + MI)	86.13%	86.16%
KNN (lexicon + MI)	85.32%	85.03%
CNN (lexicon + MI)	**89.80%**	**90.00%**
LSTM (lexicon + MI)	87.40%	87.30%

To further illustrate the effectiveness of the integration of mutual information and the domain-specific sentiment lexicon at significantly improving the performance of the classifiers, the CNN classifier, which outperforms the other classifiers in each metric, is selected for analysis. Table 10 compares the performance of the variants of the CNN classifier integrated with the sentiment lexicon alone, mutual information alone, and both the sentiment lexicon and MI.

Analysis of Table 10 reveals the following. The addition of the sentiment lexicon improves the precision, recall, and F1-score for the CNN classifier by 2.92%, 2.9%, and

Table 10. Performance of the CNN classifier in each condition

Classifier	Precision	Recall	F1-score	Macro-aveage	Micro-aveage
CNN	80.20%	80.10%	80.4%	80.30%	80.20%
CNN (lexicon)	83.12%	83.00%	83.34%	83.10%	83.40%
CNN (MI)	86.10%	85.23%	85.26%	85.41%	85.22%
CNN (MI + lexicon)	**91.6%**	**91.53%**	**91.53%**	**91.44%**	**91.53%**

2.94%, respectively, and increases its macro- and microaverages by 2–3%. The addition of MI appears to be more effective at improving the performance of the CNN classifier. Specifically, this addition increases the precision, recall, and F1-score for the CNN classifier by 5.9%, 4.93%, and 4.86%, respectively, and its macro- and microaverages by 5–6%. These results show that the method introduced in this study is remarkably effective at improving the performance of classifiers, lending credence to its validity.

6 Conclusions

In this paper, we proposed association rules and word2vec algorithm to extract aspect words of policy reviews, and based on the principle of context nearest neighbor, we extracted evaluation words and aspect-level binary groups of policy reviews. Furthermore, mutual information is introduced to expand the sentiment dictionary in policy opinion domain based on SO-PMI. This makes the model to work well on specific domain in aspect-based sentiment analysis. We presented the experiments using four machine learning models and two deep learning models, and showed that CNN with mutual information and domain sentiment dictionary has the best classification performance. The results of this paper have important decision-making value for the government to better understand the public's demand for policy. For example, in the early stage of policy release stage, government administrators can detect how the public react to the policy and what aspects the citizens discuss. Then, government administrators could adjust the policy based on the opinion concerns with the specific policy elements. Moreover, the aspect-based emotion analysis framework designed in this paper can be extended to other research fields in the future.

Acknowledgment. The authors acknowledge financial support from the National Natural Science Foundation of Beijing (No. 9222030), National Natural Science Foundation of China (No. 71774154, 72074205), Key Project of the Education Department of Jilin Province (No. JJKH20221012KJ).

References

1. Burstein, P.: The impact of public opinion on public policy: a review and an agenda. Polit. Res. Q. **56**(1), 29–40. Sage Publications, Thousand Oaks, CA (2003)

2. Anne, R., Stefanie, R., Dimiter, T.: The opinion-policy nexus in Europe and the role of political institutions. Eur. J. Polit. Res. **58**(2), 412–434. Blackwell Publishing Ltd. (2019)
3. Belkahla, O., Mellouli, S., Trabelsi, Z.: From citizens to government policy-makers: social media data analysis. Gov. Inf. Q. **36**(3), 560–570 (2019)
4. Xiong, J., Feng, X., Tang, Z.: Understanding user-to-User interaction on government microblogs: an exponential random graph model with the homophily and emotional effect. Inf. Process. Manag. **57**(4), 102229 (2020)
5. Hagen, L.: Content analysis of e-petitions with topic modelling: how to train and evaluate LDA models? Inf. Process. Manag. **54**(6), 1292–1307 (2018)
6. Driss, O.B., Mellouli, S., Trabelsi, Z.: From citizens to government policy-makers: social media data analysis. Gov. Inf. Q. **36**(3), 560–570 (2019)
7. Depaula, N., Dincelli, E., Harrison, T.M.: Toward a typology of government social media communication: democratic goals, symbolic acts and self-presentation. Gov. Inf. Q. **35**(1), 98–108 (2018)
8. Sobkowicz, P., Kaschesky, M., Bouchard, G.: Opinion mining in social media: modelling, simulating, and forecasting political opinions in the web. Gov. Inf. Q. **29**(4), 470–479 (2012)
9. Pang, B., Lee, L., Vaithyanathan, S.: Thumbs up? Sentiment classification using machine learning techniques. In: Proceedings of the 2002 Conference on Empirical Methods in Natural Language Processing ({EMNLP} 2002), pp. 79–86. Association for Computational Linguistics (2002)
10. Schouten, K., Frasincar, F.: Survey on aspect-level sentiment analysis. IEEE Trans. Knowl. Data Eng. **28**(3), 813–830 (2015)
11. Jakob, N., Gurevych, I.: Extracting opinion targets in a single and cross-domain setting with conditional random fields. In: Proceedings of the 2010 Conference on Empirical Methods in Natural Language Processing, pp. 1035–1045 (2010)
12. Zhang, S., et al.: Opinion analysis of product reviews. In: Proceedings of the 2009 Sixth International Conference on Fuzzy Systems and Knowledge Discovery, vol. 2, pp. 591–595 (2009)
13. Poria, S., et al.: Sentic LDA: improving on LDA with semantic similarity for aspect-based sentiment analysis. In: Proceedings of the 2016 International Joint Conference on Neural Networks (IJCNN), pp. 4465–4473 (2016)
14. Hofmann, T.: Probabilistic latent semantic indexing. In: Proceedings of the 22nd Annual International ACM SIGIR Conference on Research and Development in Information Retrieval, pp. 50–57 (1999)
15. Hu, M., Liu, B.: Mining and summarizing customer reviews. In: Proceedings of the Tenth ACM SIGKDD International Conference on Knowledge Discovery and Data Mining, pp. 168–177 (2004)
16. Zhu, Y.-L., et al.: Semantic orientation computing based on HowNet. J. Chin. Inf. Process. **20**(1), 14–20 (2006)
17. Turney, P.D., Littman, M.L.: Measuring praise and criticism: Inference of semantic orientation from association. ACM Trans. Inf. Syst. (TOIS) **21**(4), 315–346. ACM New York, NY, USA (2003)
18. Lin, C., He, Y., Everson, R., et al.: Weakly supervised joint sentiment-topic detection from text. IEEE Trans. Knowl. Data Eng. **24**(6), 1134–1145 (2012)
19. Duric, A., Song, F.: Feature selection for sentiment analysis based on content and syntax models. Decis. Support Syst. **53**(4), 704–711 (2012)
20. Huang, Z., Xu, W., Yu, K.: Bidirectional LSTM-CRF models for sequence tagging (2015). arXiv preprint arXiv:1508.01991
21. Behera, R.K., Jena, M., Rath, S.K., et al.: Co-LSTM: convolutional LSTM model for sentiment analysis in social big data. Inf. Process. Manag. **58**(1), 102435 (2021)

Estimation of Network Efficiency Based on Sampling

Hongyu Dong and Haoxiang Xia(✉)

Institute of Systems Engineering, Dalian University of Technology, Dalian 116024, Liaoning, China
hxxia@dlut.edu.cn
http://faculty.dlut.edu.cn/hxxia/zh_CN/index.htm

Abstract. Network efficiency can characterize the cost of information exchange in the network. It is widely used as an important indicator, but its calculation is very complex and time-consuming. In this paper, a sampling-based global efficiency estimation method is proposed. The global efficiency of the network is estimated using the statistical characteristics of the shortest path length distribution and node efficiency of the network. Different strategies for selecting nodes are compared. During the sampling process, the normality of the distribution of node efficiency is checked. And an adaptive method for adjusting algorithm parameters is proposed. The test results show that the method can greatly reduce the calculation time on the premise of ensuring the accuracy, and can maintain an appropriate balance between accuracy and speed.

Keywords: Complex network · Global efficiency · Sampling

1 Introduction

A complex network is a network structure composed of a large number of nodes and edges [1], which can represent many real-world complex systems. Different nodes in the network usually have unique importance according to background context, which can be measured by centrality [2] and other properties.

In network research, it is usually necessary to measure the efficiency of information exchange, for which Latora V et al. proposed the concept of network efficiency [3]. Network efficiency combines average distance and clustering coefficient, which can better describe the global and local information transfer efficiency. It is widely applied in many research fields.

Although widely used, the process of computing network efficiency is very time-consuming and complex. The computation time complexity in undirected graphs can reach to $O(n^3)$ [4] (where n is the number of nodes in the network). The computational complexity greatly limits the application of network efficiency in evaluating large-scale networks. There have been many researches on approximation of network centrality, but estimation method of network efficiency is rarely reported.

J. Chen et al. (Eds.): KSS 2022, CCIS 1592, pp. 218–225, 2022.
https://doi.org/10.1007/978-981-19-3610-4_16

In this paper, an estimation algorithm based on sampling is proposed based on the characteristics of long computation time and high complexity for the global efficiency of the network. By examining the distribution of the average path length in the network and evaluating its normality, the influence of different node extraction strategies on the estimation accuracy is deeply analyzed, and the experimental results are analyzed and discussed.

2 Definations and Notations

In this paper, network is represented by $G = (V, E)$, where $V = \{v_1, v_2, \ldots, v_n\}$ represents the node set of the network G, and $E = \{e_1, e_2, \ldots, e_n\}$ represents the edge set.

The node efficiency E_i refers to the difficulty of connecting node i to all other nodes in the network, and is numerically equal to the harmonic average of the distances from node i to all other nodes, namely

$$E_i = \frac{1}{n-1} \sum_{j=1, j \neq i}^{n} \frac{1}{d_{ij}} \tag{1}$$

The efficiency calculation of the whole network can be carried out at global and local levels respectively. The global efficiency characterizes the ability of the entire network to exchange and transmit information at the same time, and is numerically equal to the average of all node efficiency, namely

$$E_{glob} = \frac{1}{n(n-1)} \sum_{i \neq j \in G}^{n} \frac{1}{d_{ij}} = \frac{1}{n} \sum_{i=1}^{n} E_i \tag{2}$$

3 Estimation Algorithm

3.1 Shortest Path Length Estimation

It can be seen from the definition of the global efficiency that it is essentially the harmonic average of all paths between nodes in the network. For a more complex network, as the number of nodes increases, the number of paths between nodes also increases rapidly. Research found that some intermediate nodes often appear in the paths of other node pairs, and some rarely appear in paths. That is, probability distribution of nodes appearing in all shortest paths is not uniform [5]. Therefore, there is a natural idea: if it is possible to study the distribution law of the shortest path in the network, and use some sampling method to replace the overall distribution with sampled ones, and then calculate the global efficiency of the network.

There are few studies on the shortest path length distribution in the network. Katzav E et al. studied the shortest path length distribution of ER random network, gave an approximate formula and compared it with the actual generative network, showing good fitting results [6]. Ventrella A V et al. analyzed and modeled the shortest path distribution, and believed that the distribution in scale-free network showed characteristics of Gaussian distribution [7]. After observing and analyzing a large number of rule-generated networks and real-world networks, Ye Q et al. believed that the shortest path lengths of complex networks showed normal distributions with different mean and skewness [8].

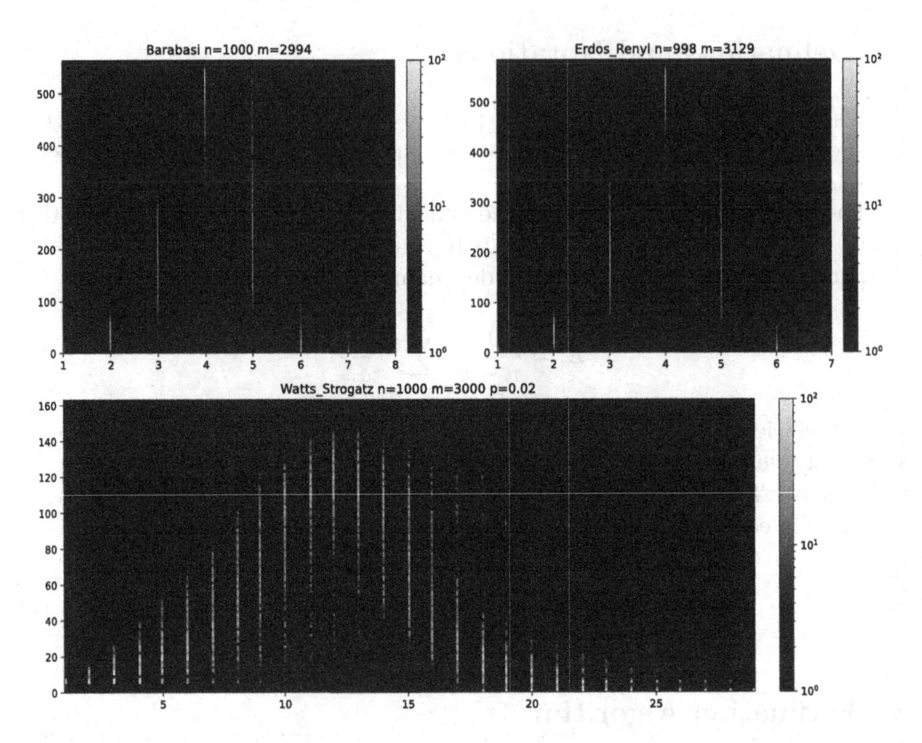

Fig. 1. Heat map of shortest path length distribution of random networks

Figure 1 shows distribution of shortest path lengths in 3 types of networks (a: WS network, n = 1000, m = 4000, p = 0.0; b: BA network, n = 1000, m = 4000; c: ER network, n = 1000, m = 4000), the path length from each node to all other nodes is represented by a line graph, the horizontal axis is the path length, and the vertical axis is the count corresponding to the path length. The right side is the legend, and the color indicates the distance distribution range of each node to other nodes.

3.2 Sampling Method

According to Hoeffding's inequality [9], the sample mean and expected error of any bounded random variable can be bounded within any constant range. The efficiency of one node can be regarded as an independent random variable. Based on the given accuracy ξ, we can limit the path distribution error of sampled nodes within a certain range, thereby replacing all shortest path distributions of the network with randomly chosen nodes.

According to Chebyshev's law of large numbers, the mean of mutually independent random variables converges to the mathematical expectation of the variable according to probability. As the number of samples increases, the average value of random variables will approach a constant i.e. the global efficiency of the network. The node efficiency is calculated separately for the selected nodes, and the efficiency value of each node can be regarded as one independent random variable. This algorithm averages the efficiencies of all sampling nodes as the estimation result of the global efficiency of the network.

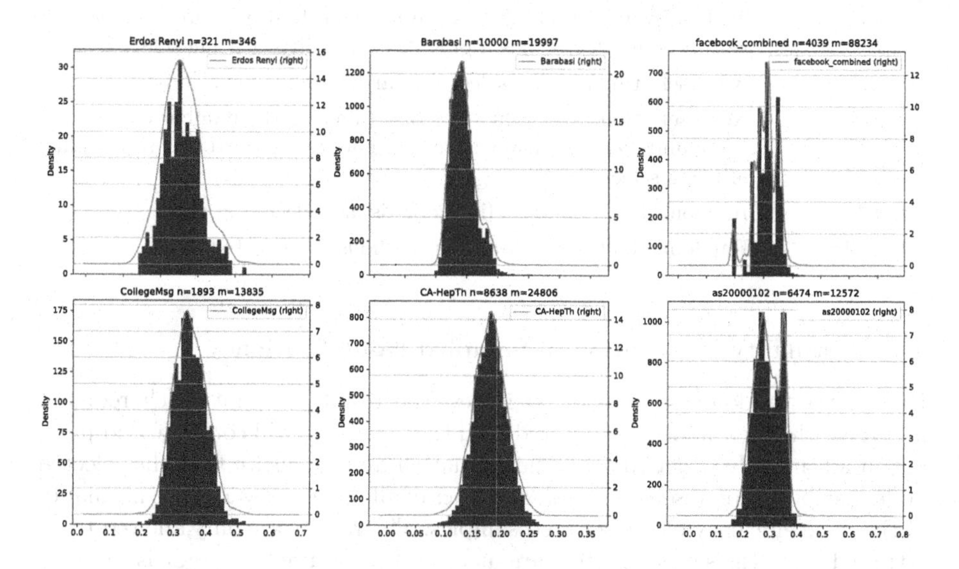

Fig. 2. Histogram of node efficiency distribution in some networks

Figure 2 shows the histogram of the node efficiency distribution of 6 networks of different scales. The horizontal axis is the node efficiency and the vertical axis is the number of nodes corresponding to the efficiency, and the red dotted line is the fitted normal distribution curve. It can be seen from the figure that the node efficiency presents an approximate normal distribution, and the larger the number of nodes in the network, the better the normality. However, when the number of sampling nodes is large enough, the conditions for using Chebyshev's law of large numbers are met.

3.3 Node Selection Strategy

Under the premise of a certain number of selected nodes, how can the selected nodes be distributed as evenly as possible in the network? Or make the selected nodes representatively stand for different types of nodes in the network. This leads us to think about strategy of how to select nodes. Brandes U et al. proposed several strategies for source node selection in algorithm for estimating betweenness centrality, such as RanDeg, MaxMin, Mixed, etc. [10]. This algorithm adopts several of them and proposes one new strategy of selecting nodes according to community division. As described in the following table (Table 1):

Table 1. Node selection strategy

Strategy Name	Description
Random	Randomly select nodes
Randeg	Node selection probability is proportional to degree, and the node with a larger degree has a greater probability of being selected
MaxMin	After selecting a node, select the farthest node next
RanComm	Node selection probability is proportional to the number of its community. community with a more nodes has a greater probability of being selected
Mixed	Random, Randeg, MaxMin methods are selected in turn
Mixed2	Random, RanComm, MaxMin methods are selected in turn

3.4 Normality Verification of Sampled Node Efficiency

Due to the randomness of selected nodes, the result may not reach required accuracy when few nodes are selected, and the algorithm will converge and jump out in advance. In order to avoid this situation and to ensure that the selected nodes can better represent the characteristics of all nodes, we verify the normality of the node efficiency of the selected samples. The results are only allowed to be returned when the set parameters are met and the required accuracy is reached.

This algorithm performs the Kolmogorov-Smirnov test on the efficiency of the selected sample nodes. When the D statistic of the distribution passes the hypothesis test at a given significance level, the distribution is considered to be a normal distribution.

3.5 Adaptive Adjustment of Algorithm Parameters

When desired accuracy is not given, algorithm needs to keep a reasonable balance between sampling accuracy and computation time. This algorithm mainly adapts to various calculation accuracy by adjusting the number of nodes selected in each batch and convergence value. The number of nodes selected each time mainly determines the total calculation time of the algorithm, and the allowable convergence accuracy mainly determines the accuracy of the calculation result. After a lot of experiments, the default value of the algorithm is set to: select 0.1% of the total number of nodes in the network each time (10 if the network size is too small), and the initial convergence value is 0.1% of the first calculated value (If global efficiency still does not converge after a certain number of steps, it is set to 0.001).

After setting the initial value, algorithm will dynamically adjust the number of nodes and the convergence value according to the calculation results of the previous round. Set the network efficiency value of each round of calculation as E_i, the number of extracted nodes as n_i, and set the convergence accuracy as ε_i. E is the network efficiency column calculated in each round, and the average value of each round is recorded as μ_i, and the standard deviation is σ_i. Let the constants $P = 0.95$ and $Q = 1.01$. Update the parameters as follows:

$$n_{i+1} = n_i P \left| \frac{E_i - \mu_i}{\sigma_i} \right| \tag{3}$$

$$\varepsilon_{i+1} = \varepsilon_i Q \left| \frac{E_i - \mu_i}{\sigma_i} \right| \tag{4}$$

3.6 Estimation Algorithms of Network Efficiency

This algorithm performs iterative calculation, and each round of calculation process can be divided into three steps. First nodes are selected according to sampling strategy and the node efficiency is calculated. Then determine whether current result meets the ending condition. Finally end the calculation or update the parameters and continue the loop until it jumps out. The specific algorithm is described as follows:

Algorithm 1. Global Efficiency Estimation Algorithm
Input network $G = (V, E)$, node selection strategy S
(optional: number of nodes selected in a batch n, iteration accuracy ε)
Output estimated global efficiency E

1) Determine number of nodes n selected in each batch and iteration accuracy ε
2) Name extracted node list ls and number of sampled nodes a
3) Name list E_list for each round of calculation efficiency
4) initial round $i = 1$
5) **While True**:
6)   **While True**:
7)   Select n nodes from G according to stragety S
8)   Calculate efficiency of these nodes
9)   **if** selected node efficiency combined with E_list is normally distributed
10)   update ls and a
11)   **break**
12)   **if** number of sampleds is greater than network nodes **or** $|E - \mu| < \sigma$
13)   **break**
14)   **else**
15)   update n and ε
16) $i = i + 1$
17) **return** estimated global efficiency

4 Experimental Validation

Different node selection strategies are analyzed. Efficiency calculation is carried out by selecting 500 nodes fixedly in 5 networks of different sizes. Comparing different strategies, it is found that the most common random algorithm has the best accuracy, and also performs best in different scales and types of networks. Other more complex sampling strategies also have a certain time-consuming extension due to the additional index and sum operations, and do not show better accuracy than the random strategy.

In order to verify the validity and reliability of the algorithm proposed in this paper, the algorithm was tested on random networks generated by rules and real networks from SNAP (Stanford Large Network Dataset Collection) such as wikilink and p2p. The results are shown in Table 2.

Table 2. Validation of random sampling algorithm on network (20 times average)

Network	Nodes	Edges	Global efficiency	Error	Calculated time	Time saving
BA	2000	5994	0.2679	0.97%	0.1515 s	94.13%
WS 0.1	2000	6000	0.1915	0.31%	0.1541 s	96.17%
ER	2000	6037	0.2388	2.60%	0.1601 s	97.94%
2002-wiki	27526	195385	0.2571	1.20%	46.20 s	97.30%
Oregon1	10670	22002	0.2939	0.30%	4.1215 s	99.84%
CollegeMsg	1893	13835	0.3499	1.30%	0.1725 s	96.60%
p2p	36646	88303	0.181	0.70%	59.90 s	98.40%
Wiki-Vote	7066	100736	0.3267	1.02%	2.995 s	93.30%

The test results show that the method can greatly reduce the calculation time on the premise of ensuring the accuracy, and can maintain a proper balance between accuracy and speed. It has good computational accuracy for random networks as well as real-world networks, which is more pronounced on networks with low network density. Random sampling strategies show more accurate estimation results than other more complex sampling methods.

This method can reduce the computational complexity to $O(kn)$ (where k is a constant depends on desired precision). Since the efficiency calculations of each sampling node are independent to each other, the method described in this paper can perform parallel calculations on multi-core CPUs, thereby further reducing the calculation time.

5 Conclusion

In this paper we propose one estimation algorithm for global efficiency based on node sampling to overcome the disadvantages of complex network global efficiency calculation and its limited application in large-scale networks. By examining the distribution of the average path length in the network, a batch of nodes are sampled if normality meets the requirements, and the algorithm uses partial path distribution to replace the global length distribution. The accuracy of different sampling strategies is analyzed. Algorithm is verified in both artificially generated network and real network. The experimental results show the effectiveness and reliability of the algorithm, and it has good performance on networks of different categories and scales.

There are some limitations for the present work. The robustness of the proposed method still needs to be further studied. The mathematical analyses on the proposed method will improve its rigor. Further validations on more networks of different scales are also required. Finally, further research is needed on the impact of network structure on sampling accuracy and estimation time.

References

1. Albert, R., Barabási, A.L.: Statistical mechanics of complex networks. Rev. Mod. Phys. **74**(1), 47 (2002)
2. Rong, L., Guo, T., Wang, J.: Centralities of nodes in complex networks. J. Univ. Shanghai Sci. Technol. **30**(3), 227–230 (2008)
3. Latora, V., Marchiori, M.: Efficient behavior of small-world networks. Phys. Rev. Lett. **87**(19), 198701 (2001)
4. Floyd, R.W.: Algorithm 97: shortest path. Commun. ACM **5**(6), 345 (1962)
5. Tang, J., Wang, T.: Research on the approximation algorithms for the betweenness property computation on complex social networks. Comput. Eng. Sci. **30**(12), 9–10 (2008)
6. Katzav, E., et al.: Analytical results for the distribution of shortest path lengths in random networks. EPL (Europhys. Lett.) **111**(2), 26006 (2015)
7. Ventrella, A.V., Piro, G., Grieco, L.A.: On modeling shortest path length distribution in scale-free network topologies. IEEE Syst. J. **12**(4), 3869–3872 (2018)
8. Ye, Q., Wu, B., Wang, B.: Distance distribution and average shortest path length estimation in real-world networks. In: Cao, L., Feng, Y., Zhong, J. (eds.) ADMA 2010. LNCS (LNAI), vol. 6440, pp. 322–333. Springer, Heidelberg (2010). https://doi.org/10.1007/978-3-642-17316-5_32
9. Hoeffding, W.: Probability inequalities for sums of bounded random variables. In: Fisher, N.I., Sen, P.K. (eds.) The Collected Works of Wassily Hoeffding. Springer Series in Statistics, pp. 409–426. Springer, New York, NY (1994). https://doi.org/10.1007/978-1-4612-0865-5_26
10. Brandes, U., Pich, C.: Centrality estimation in large networks. Int. J. Bifurc. Chaos **17**(07), 2303–2318 (2007)

Research on the Maturity Evaluation of the Public Health Emergency Response Capability of Urban Communities

Mingjia Cui(✉) and Xingpeng Wang

Shijiazhuang TieDao University, Shaoxing 050000, Shijiazhuang, China
15630179656@163.com, xingpengwang@163.com

Abstract. Urban communities are the basic units for preventing and responding to public health emergencies; thus, it is of great significance to scientifically evaluate the public health emergency capability of urban communities. This article establishes a community public health emergency response capability evaluation index system from the four aspects of prevention, preparation, response, and recovery and uses the CRITIC method to determine the index weights. This article draws on the idea of the capability maturity of the software development process, constructs an urban community public health emergency maturity model, analyzes the maturity characteristics of each level of capability, and combines the cloud model to evaluate the maturity level of the community public health emergency response capability. Finally, a typical Shijiazhuang community is used as the research object to evaluate the emergency response capacity, analyze the evaluation results and put forward targeted suggestions. The goal is to enrich and develop the management system of public safety incidents in urban communities in China.

Keywords: Public health emergency · Capability maturity level · Cloud model evaluation

1 Introduction

Urban communities are the direct subject that responds to public health emergencies. The outbreak of the 2019 coronavirus disease (COVID-19) in early 2020 revealed that most urban communities in China are still vulnerable to public health emergencies and that their emergency management capability needs to be improved [1]. Due to the weak emergency management capability of the community itself, the lack of a comprehensive emergency management mechanism, and the weak emergency awareness of community residents, the mode of responding to public health emergencies, with the community as the unit, is not yet mature. Therefore, to respond to public health emergencies, establishing a systematic and scientific evaluation system for urban communities holds great significance [2]. At present, most capability evaluation research index systems focus on hardware research, such as infrastructure, and on static evaluation. The concept of process-based dynamic evaluation is still not mature, while the concept and ideas of hierarchical management and the continuous improvement of the software development

J. Chen et al. (Eds.): KSS 2022, CCIS 1592, pp. 226–241, 2022.
https://doi.org/10.1007/978-981-19-3610-4_17

process capability maturity model fit perfectly with the management of public health emergencies in urban communities in China.

Scholars have performed research and practice on the emergency management capability of urban communities in many fields and have obtained rich results. From the perspective of application fields, these studies and practices mainly involve typhoons [3–5], earthquakes [6, 7], floods [8–10] and other natural disasters and climate change [11–13]. However, there are relatively few studies on public health emergencies that result from man-made causes, and the research results are not comprehensive or systematic. At present, emergency management capability evaluation is gradually developing in the direction of dynamic assessment and a process orientation. Based on software development process capability maturity theory [14–19], this article takes epidemic prevention and control as an opportunity, classifies the ability of urban communities to respond to public health emergencies, builds an index evaluation system [20, 21], and combines the cloud model evaluation method to evaluate the maturity level of the emergency response capacity for public health emergencies in typical communities in Shijiazhuang. It provides a reference for the transformation and upgrading of urban community functions in China, promotes the scientificization of urban community emergency management systems, and improves the crisis response capability of communities.

2 Model of the Maturity Evaluation of the Public Health Emergency Response Capability of Urban Communities

2.1 Connotation of the Capability Maturity Model

The Software Engineering Institute (SEI) of Carnegie Mellon University in the United States first proposed the capability maturity model (CMM) from the perspective of software process capability. Later, many scholars applied CMM ideas and concepts to various fields. The CMM provides a multistep evolutionary framework for the realization of the goals of an organization. The maturity model shows a dynamic optimization idea of continuous evolution and continuous improvement. The core of the CMM is to evaluate the maturity level of an organization's object, find areas for improvement based on the level results, and form a continuous improvement model. By using CMM to evaluate the emergency response capability of public health emergencies in urban communities, the maturity level of emergency response capabilities can be judged, and weaknesses can be found to realize the scientific management of public health emergencies in urban communities.

2.2 The Architecture of the Emergency Capability Maturity Model

Through an analysis of indicators and the description of each level of the maturity model, the capability of an urban community to respond to public health emergencies is longitudinally evaluated. From a horizontal perspective, the CMM defines the evaluation standards and guidelines for each level. Based on CMM theory, this paper divides the capability of urban communities to respond to public health emergencies into five levels. The five-level maturity model framework is shown in Fig. 1. When the maturity level is higher, the capability of urban communities to manage and control public health emergencies is stronger, and they can more easily reduce losses and ensure safety and stability.

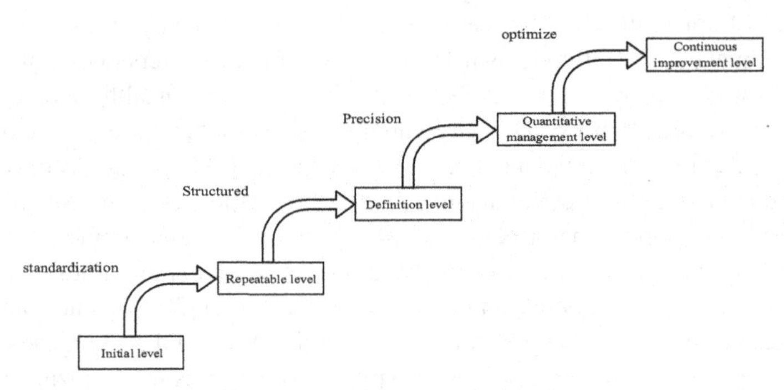

Fig. 1. Capability maturity levels.

Table 1 shows a description of the maturity characteristics of each level of urban community public health emergency response capability.

Table 1. Description of the maturity levels of the public health emergency response capability of urban communities.

Maturity level	Grade description
Initial level	Community public health emergency management is imperfect, professional team management is lacking, the emergency response mechanism is not perfect, the risks are unpredictable, and emergency work is full of chaos
Repeatable level	The community has initially established an emergency management mechanism and emergency management organization, there are relatively complete medical facilities, rules and regulations have been gradually improved, and hierarchical and classified risk management has been realized
Definition level	The community emergency plan system is gradually being completed, public health emergency training and drills are effectively implemented, the monitoring and early warning capabilities are improved, and the emergency medical security conditions are sound

(*continued*)

Table 1. *(continued)*

Maturity level	Grade description
Quantitative management level	The community rescue team is professional, with a flexible emergency organization and coordination capability, the community has a stable emergency management system for data management, and there is a sound information communication mechanism
Continuous improvement level	The community has established a complete incident investigation and accountability system that can provide community residents with life assistance and psychological counseling, and the community has a complete reconstruction plan and a complete mechanism for subsequent learning and improvement

2.3 The Internal Structure of the Emergency Capability Maturity Model

To better study the public health emergency response capability of communities, the key area (KA) concept, which more completely explains the key features of each level of the CMM, is introduced. Each KA includes a key process area (KPA). A KPA describes the problems faced by each maturity level and reduces these problems to development goals. Each KPA includes key practices (KPs), that is, specific methods for achieving a series goals. The internal structure of the model is shown in Fig. 2.

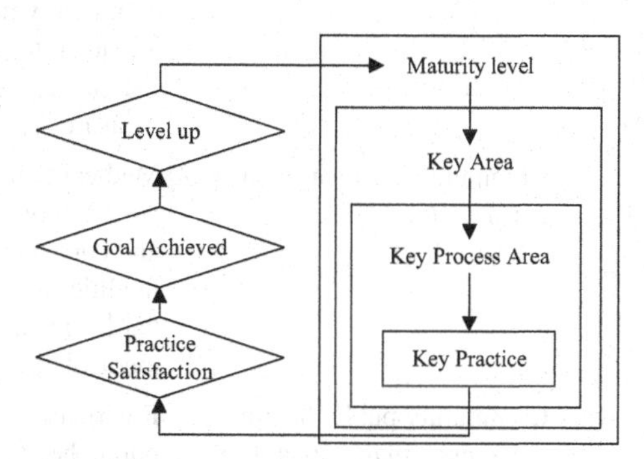

Fig. 2. The internal structure of the maturity level of the public health emergency response capability of urban communities.

3 Maturity Rating Model

3.1 Establishment of the Indicator System

Based on the development stage of public health emergencies, the public health emergency management capability of urban communities is divided into prevention capability, preparation capability, response capability, and recovery capability. Based on these four stages, an indicator system is established. Each capability has four indicators of different maturity levels that correspond to all of them except for the initial level. The specific indicators are shown in Table 2 below.

Table 2. Index system of the public health emergency response capability of urban communities.

Primary indicators	Secondary indicators	Indicator description
Prevention capability (A)	Community public health knowledge publicity and education (A1)	Community staff use various forms to publicize public health knowledge to community residents; staff members regularly make return visits to assess whether publicity and education are in place
	Community public health emergency rule and regulation construction (A2)	Community staff has been organized to establish the formulation, implementation, feedback and guarantee of community-based public health emergency regulations
	Community public health risk identification and evaluation (A3)	Community managers identify community public health risks, these risks are classified by size, community public health risks are quantitatively evaluated, and risk prevention and control measures are formulated
	Community medical facility construction (A4)	Medical clinics in the community and the equipment and consumption of medical facilities are established; as a hardware guarantee, the relevant medical equipment is improved
Preparation capability (B)	Community public health emergency plan preparation (B1)	Emergency plans for community public health emergencies are prepared, community management personnel organize professionals to evaluate the feasibility of the emergency plans, and the emergency plans are improved and optimized

(continued)

Table 2. (*continued*)

Primary indicators	Secondary indicators	Indicator description
	Community public health emergency training and drills (B2)	Community managers organize community residents to conduct public health emergency training and drills in advance
	Community epidemic situation monitoring and early warning capability (B3)	A procedural public health monitoring and warning mechanism is established, the public health warning equipment is equipped to sensibly respond to public health emergencies, and communities use information technology for monitoring during an epidemic
	Community public health emergency medical insurance (B4)	Community public health emergency funds are established and managed, a community-wide medical assistance team is formed, medical staff regularly participate in evaluation and training, and there are emergency supplies and guarantees
Response capability(C)	Community rescue team construction (C1)	A public health community rescue team is built, professional rescue guidance is provided to the emergency rescue team, and the rescue team is scientifically and rationally divided
	Community emergency organization and coordination capability (C2)	Community managers have the ability to organize community residents, volunteers, and medical staff and to coordinate equipment, funds and relief supplies
	Community public health emergency management system response capability (C3)	Information technologies such as the Internet of things and cloud computing are used to establish a public health emergency management information system, the safety and stability of the system are improved, and the efficiency of the system in collecting and analyzing community epidemic data is improved

(*continued*)

Table 2. (*continued*)

Primary indicators	Secondary indicators	Indicator description
	Community public health incident information communication ability (C4)	Each subject has the ability to transmit and communicate effective information obtained during an outbreak
Recovery capability (D)	Investigation, evaluation and accountability of community public health incidents (D1)	After the emergency, there is an investigation into its cause, and there is accountability; the severity of the public health emergency is assessed
	Life assistance and psychological counseling for community residents (D2)	Residents are compensated for the life and economic losses caused by public health emergencies; the psychological shadow of community residents is promptly relieved
	Community reconstruction plan and evaluation (D3)	After the public health emergency, community managers are organized to formulate a community reconstruction plan; the community plan is evaluated and implemented
	Subsequent learning and improvement mechanism (D4)	After public health emergencies, the ability to learn is improved; a continuous improvement learning mechanism is established

3.2 Key Process Areas of Each Maturity Level

Based on the construction of the indicator system, the system is combined with maturity model as the main content of the KPAs. The division of the KPAs for each maturity level is shown in Fig. 3.

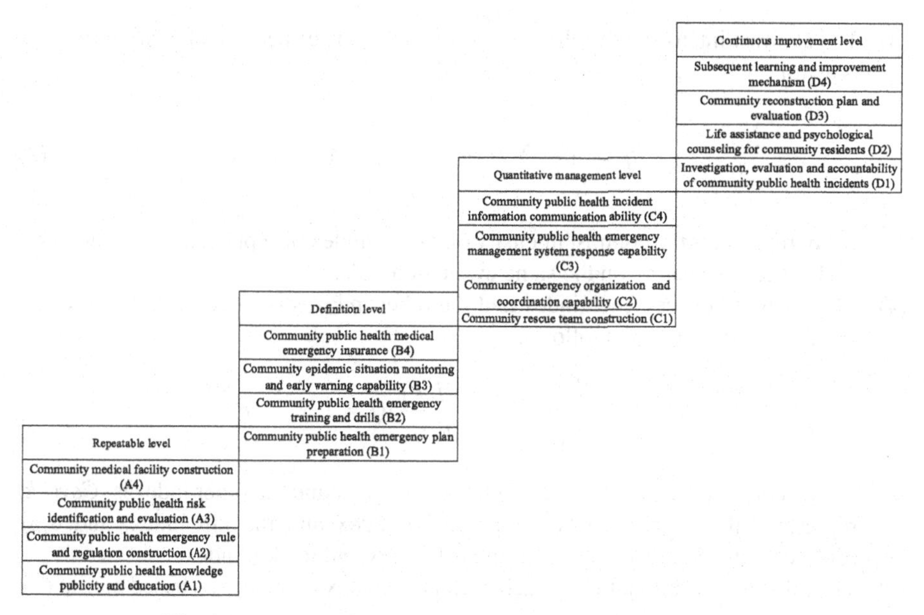

Fig. 3. Division of the key process areas of the maturity model.

3.3 Evaluation Steps and Methods

This paper uses the cloud model to evaluate the maturity of the public health emergency response capability of urban communities and uses the CRiteria Importance Through Intercriteria Correlation (CRITIC) method proposed by Diakoulaki to determine the index weights.

Determination of the Index Weights

Taking into account the volatility and correlation of the indicators, this article uses the CRITIC method to determine the index weights. The CRITIC method describes the weight of the evaluation index through the two concepts of contrast intensity and conflict. Contrast intensity describes the evaluation value gap within the same index, and conflict is described by the correlation between the indicators. If the indicators show a positive correlation, then conflict is not high. The specific calculation steps are as follows.

(1) An index system and index evaluation standards for the construction of the public health emergency response capability of urban communities are established.
(2) An index evaluation matrix is established, and the evaluation results of m evaluators on n indexes are expressed as matrix R.

$$R = \begin{bmatrix} x_{11} \ldots x_{1j} \ldots x_{1n} \\ x_{i1} \ldots x_{ij} \ldots x_{in} \\ x_{m1} \ldots x_{mj} \ldots x_{mn} \end{bmatrix} (\mathrm{i} = 1,2,\ldots\mathrm{m}\ ;\mathrm{j} = 1,2,\ldots,\mathrm{n}) \quad (1)$$

x_{ij} represents the evaluation value of the i-th evaluator on the j-th index, m represents the number of evaluators and n represents the number of indexes.

(3) The contrast intensity of each index is calculated. The specific calculation formula is as follows:

$$\sigma_j = \sqrt{\frac{1}{m}\sum_{i=1}^{m}(x_{ij} - \mu)^2} (j = 1, 2, ..., n) \tag{2}$$

σ_j represents the contrast intensity of the j-th index, μ represents the mean value of different evaluators on the same evaluation index.

(4) The conflict between each index and the other indexes is calculated. The specific calculation formula is as follows:

$$y_j = \sum_{k=1}^{n}(1 - \frac{Cov(j,k)}{\sigma_j\sigma_k})(j = 1, 2, ..., n) \tag{3}$$

y_j represents the conflict between the j-th index and the other indexes, *Cov(j,k)* represents the covariance between the jth index and the kth index, and $\sigma_j\sigma_k$ represents the contrast strength of the j-th index and the kth index.

(5) The information amount C_j of each evaluation index is calculated based on Eq. (4).

$$C_j = \sigma_j y_j (j = 1, 2, ..., n) \tag{4}$$

(6) According to the information amount, the objective weight W_j of the indicator is obtained.

$$W_j = \frac{C_j}{\sum_{j=1}^{n} C_j}(j = 1, 2, ..., n) \tag{5}$$

According to the above calculation steps, the index weights of all levels are shown in Table 3.

Table 3. Index weights of the emergency response capacity for public health emergencies in urban communities.

Primary indicators	Index weight	Secondary indicators	Index weight
Prevention capability (A)	0.273	Community public health knowledge publicity and education (A1)	0.244
		Community public health emergency rule and regulation construction (A2)	0.262

(continued)

Table 3. (*continued*)

Primary indicators	Index weight	Secondary indicators	Index weight
		Community public health risk identification and evaluation (A3)	0.237
		Community medical facility construction (A4)	0.257
Preparation capability (B)	0.256	Community public health emergency plan preparation (B1)	0.237
		Community public health emergency training and drills (B2)	0.269
		Community epidemic situation monitoring and early warning capability (B3)	0.251
		Community public health emergency medical insurance (B4)	0.243
Response capability(C)	0.290	Community rescue team construction (C1)	0.231
		Community emergency organization and coordination capability (C2)	0.294
		Community public health emergency management system response capability (C3)	0.225
		Community public health incident information communication ability (C4)	0.250
Recovery capability (D)	0.181	Investigation, evaluation and accountability of community public health incidents (D1)	0.262

(*continued*)

Table 3. (*continued*)

Primary indicators	Index weight	Secondary indicators	Index weight
		Life assistance and psychological counseling for community residents (D2)	0.253
		Community reconstruction plan and evaluation (D3)	0.235
		Subsequent learning and improvement mechanism (D4)	0.250

Cloud Model Theory

The cloud model was proposed by Li Deyi, an academician of the Chinese Academy of Engineering. It is an uncertain conversion model that addresses qualitative concepts and quantitative descriptions. It has been widely used in decision analysis, intelligent control, image processing and other fields. The cloud model uses three numerical characteristics (*Ex*, *En*, *He*) to describe the evaluation of qualitative indicators, *Ex* represents the expected distribution of the indicators, and entropy *En* determines the ambiguity and randomness of the qualitative indicators, which is shown in the figure as the span of the sample cloud. When the concept is larger, it is more macroscopic and fuzzier. Superentropy *He* is the "entropy of entropy", which describes the degree of dispersion of entropy and reflects the cohesion of the degree to which each evaluation value belongs to this index value. Through the digital features of the cloud model, a normal distribution cloud map can be generated to visually display the evaluation results, as shown in Fig. 4.

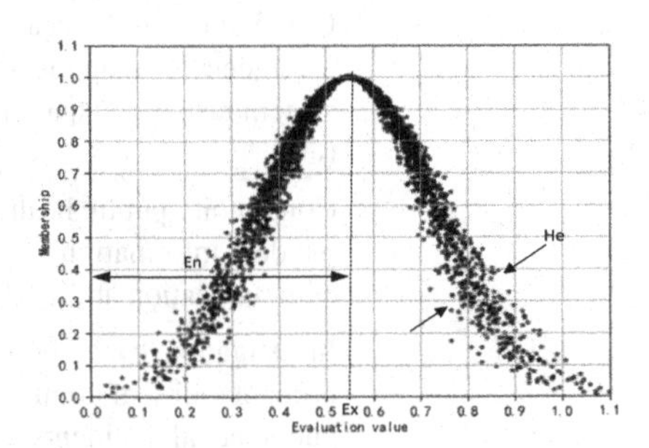

Fig. 4. Digital feature map of the cloud model.

Comprehensive Evaluation Steps of the Cloud Model

(1) Calculation of the digital characteristics of the index clouds at each level

The inverse cloud generator is used to obtain the cloud model digital features (Exj, Enj, Hej) of each index of the m evaluation values xi (i = 1, 2., m) of the jth index. Their calculation formulas are shown in Eqs. (6), (7), and (8).

$$Ex_j = \frac{1}{m}\sum_{i=1}^{m} x_i, j = 1, 2, ..., n \tag{6}$$

$$En_{\mathrm{j}} = \sqrt{\frac{\pi}{2}}\frac{1}{\mathrm{m}}\sum_{i=1}^{m} |x_i - Ex_j|, j = 1, 2, ..., n \tag{7}$$

$$He_{\mathrm{j}} = \sqrt{\frac{1}{m-1}\sum_{i=1}^{m} (x_i - Ex_j)^2 - En_j^2}, j = 1, 2, ..., n \tag{8}$$

(2) Computation of comprehensive cloud digital features

The index of comprehensive cloud digital characteristics is calculated through the cloud digital characteristics of each index, and the calculation formula is shown in Eqs. (9), (10), and (11).

$$Ex = \frac{\sum_{j=1}^{n} Ex_j w_j}{\sum_{\mathrm{j}=1}^{n} w_j}, j = 1, 2, ..., n \tag{9}$$

$$En = \frac{\sum_{j=1}^{n} En_j w_j^2}{\sum_{j=1}^{n} w_j^2}, j = 1, 2, ..., n \tag{10}$$

$$He = \frac{\sum_{j=1}^{n} He_j w_j^2}{\sum_{j=1}^{n} w_j^2}, j = 1, 2, ..., n \tag{11}$$

(3) Determination of the evaluation result set

In this paper, the capability maturity level is used as the evaluation level.

$$E = (e_1, e_2, e_3, e_4, e_5) \tag{12}$$

e1 represents the initial level, e2 represents the repeatable level, e3 represents the defined level, e4 represents the quantitative management level, and e5 represents the continuous improvement level.

When the evaluation result is in the interval [0.8–1], it belongs to the continuous improvement level. When the evaluation result is in the interval [0.6–0.8], it belongs to the quantitative management level. When the evaluation result is in the interval [0.4–0.6], it belongs to the defined level. When the evaluation result is in the interval [0.2–0.4], it belongs to the repeatable level, and when the evaluation result is in the interval [0–0.2], it belongs to the initial level. Based on Eqs. (13), (14), and (15), the corresponding evaluation interval is converted into the standard digital features of the cloud model, as shown in Table 4 below, and the generated evaluation standard cloud model diagram is shown in Fig. 5 below.

$$Ex = (c_{\min} + c_{\max})/2 \tag{13}$$

$$En = (c_{\max} - c_{\min})/6 \tag{14}$$

$$He = k \tag{15}$$

x_{min} and x_{max} are the upper and lower limits of the score, respectively, and k is a constant. This article takes 0.02 as the value of k.

Table 4. Digital characteristics of the standard cloud models

Evaluation result set	Initial level	Repeatable level	Definition level	Quantitative management level	Continuous improvement level
Evaluation interval value	[0–0.2]	[0.2–0.4]	[0.4–0.6]	[0.6–0.8]	[0.8–1.0]
Normal cloud digital features	(0.1,0.033, 0.02)	(0.3,0.033, 0.02)	(0.5,0.033, 0.02)	(0.7,0.033, 0.02)	(0.9,0.033, 0.02)

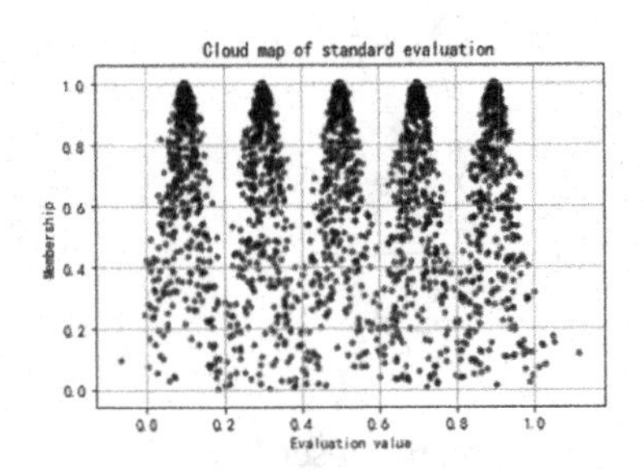

Fig. 5. Standard cloud diagram of the evaluation indicators.

(4) Generation of the cloud diagrams of each indicator

Relevant experts in the field of public health and community management workers should be invited to score the evaluation indicators of the public health emergency response capability of urban communities. Then, the reverse cloud generator is used to obtain the cloud digital characteristics of each indicator, and the corresponding indicator contrast cloud model diagram is generated.

(5) Generation of a comprehensive evaluation cloud model diagram

After obtaining each index to evaluate the cloud digital features, based on Eqs. (8), (9), and (10), comprehensive evaluation cloud digital features and cloud model diagrams of the public health emergency response capability of urban communities are obtained. Then, by comparing the standard cloud model diagram, when the standard cloud diagram cannot be accurately determined, the cloud similarity can be calculated to determine the evaluation result. A comprehensive evaluation of cloud similarity can be calculated based on Eq. (16).

$$S = \exp\left(-\frac{(Ex - Ex^*)^2}{2(En^*)^2}\right) \tag{16}$$

where Ex is the expected value of the comprehensive evaluation cloud, and Ex^* and En^* are the expected value of the standard cloud evaluation and the superentropy value, respectively. Based on the principle of maximum similarity, the evaluation level that corresponds to the maximum cloud similarity is the maturity of the public health emergency response capability of the evaluated community evaluation level. Figure 6 below indicates that the maturity level of the public health emergency response capability that corresponds to the index of the community is at the quantitative management level.

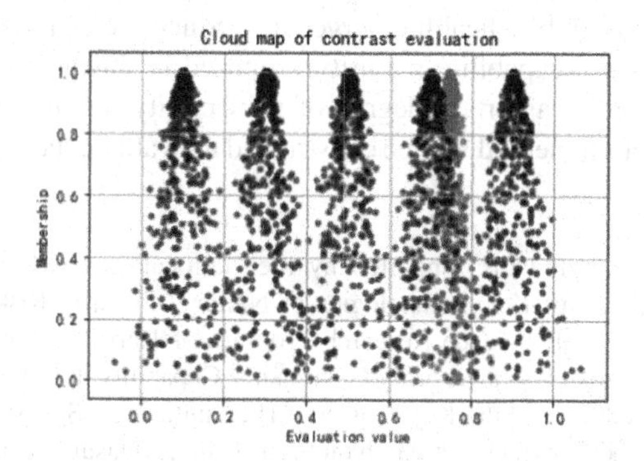

Fig. 6. Contrast cloud diagram of the evaluation indicators.

Result Analysis

As seen from Fig. 6 above, the level of emergency response capacity for public health emergencies in the typical urban communities in Shijiazhuang selected in this study is the quantitative management level. The experience and lessons of epidemic prevention and control have basically realized the controllable management of public health emergencies and proved the scientificity and accuracy of the maturity theory and cloud model. However, the ability to communicate information about community public health incidents, emergency management system data management and response capabilities, community emergency organization and coordination capabilities, and community rescue team building needs to be improved.

4 Conclusion and Discussion

By aiming at the current situation of an insufficient emergency capability to public health emergencies in urban communities. Based on the maturity theory and thought, this paper divides the community emergency capability into five levels: initial level, repeatable level, defined level, quantitative management level and continuous improvement level. And we construct a maturity model of urban community response capability for public health emergencies and use cloud model evaluation methods to evaluate the emergence of urban communities in Shijiazhuang. According to the evaluation results to determine the maturity level of community emergency response capability, so as to find the problems to be solved in community public health emergency management and put forward scientific suggestions for the improvement of urban community public health emergency response capability.

In reality, this paper combines the capability maturity theory with the whole process theory of emergency management, and constructs a systematic and comprehensive evaluation index system and capability maturity evaluation model. The model can visually display the results, has strong operability, and has guiding significance for the evaluation of urban community emergency capacity. This paper has important reference value for improving the urban community public health emergency management platform and improving the urban community emergency management ability, and has important practical significance for breaking through the risks and challenges faced by the current urban community public health emergencies. Since the evaluation index value of the emergency response capability maturity proposed in this article adopts the expert scoring method, there is a certain degree of subjectivity. In future research, a more refined index evaluation method will be constructed to evaluate the research object more accurately.

Acknowledgments. The study is supported by the National Key R&D Program of China: Research and Application of Key Technologies of National Security Risk Management (NO. 2018YFC0806900), the Soft Science Research Program of Hebei Province: Research on the Improvement Path of Urban Public Security Governance Capability in Hebei Province Driven by Big Data (NO. 20557680D) and the Key Project of Humanities and Social Sciences in Colleges and Universities of Hebei Province: Research on Performance Measurement and Optimization of Multi-agent Cooperative Network in Response to Major Emergencies (NO. SD2021067).

References

1. Duan, J., Yang, B., Zhou, L.: Planning to improve the city's immunity-a written meeting on response to the novel coronavirus pneumonia emergency. Urban Plann. **44**(02), 115–136 (2020)
2. Li, J., Liu, Y., Wu, L.: Research on community emergency response plan for public health emergencies. Chin. Primary Health Care **32**(10), 38–39 (2005)
3. Wang, M., Wang, S.: The construction and experience of resilient cities: a case study of New Orleans fighting against Hurricane Katrina. Urban Dev. Res. **25**(11), 145–150 (2018)

4. Yan, W.: Shenzhen's resilience improvement strategy from the perspective of the full-cycle disaster emergency response. In: China Urban Planning Society, Chongqing Municipal People's Government. Vibrant urban and rural beautiful human settlements-Proceedings of the 2019 China Urban Planning Annual Conference (01 Urban Safety and Disaster Prevention Planning), vol. 24, no. 8, pp. 370–377 (2019)
5. Guo, C., Sim, T., Ho, H.C.: Impact of information seeking, disaster preparedness and typhoon emergency response on perceived community resilience in Hong Kong. Int. J. Disaster Risk Reduct. **50**(6), 134–139 (2020)
6. Zhai, C., Liu, W., Xie, L.: Research progress in urban seismic toughness evaluation. J. Build. Struct. **39**(09), 1–9 (2018)
7. Hosseini, K.A., Izadkhah, Y.O.: From "Earthquake and safety" school drills to "safe school-resilient communities": a continuous attempt for promoting community-based disaster risk management in Iran. Int. J. Disaster Risk Reduct. **45**(4), 72–78 (2020)
8. Chen, C., Chen, Y., Shi, B.: Urban resilience assessment model under rain and flood disaster situation. Chin. Saf. Sci. J. **28**(04), 1–6 (2018)
9. Landström, C., Becker, M., Odoni, N.: Community modelling: a technique for enhancing local capacity to engage with flood risk management. Environ. Sci. Policy **26**(2), 8–16 (2019)
10. Noonan, D.S., Sadiq, A.-A.: Community-scale flood risk management: effects of a voluntary national program on migration and development. Ecol. Econ. **157**(4), 92–99 (2019)
11. Liu, C.: City safety, climate risk and climate adaptive city construction. J. Chongqing Univ. Technol. (Soc. Sci.) **33**(08), 21–28 (2019)
12. Zhang, M., Li, H.: Research progress on urban resilience assessment under the background of climate change. Ecol. Econ. **34**(10), 154–161 (2018)
13. Jabareen, Y.: Planning the resilient city: concepts and strategies for coping with climate change and environmental risk. Cities **31**(4), 220–229 (2013)
14. Torrecilla-Salinas, C.J., Sedeño, J., Escalona, M.J.: Agile, web engineering and capability maturity model integration: a systematic literature review. Inf. Softw. Technol. **71**(6), 92–107 (2016)
15. Lacerda, T.C., von Wangenheim, C.G.: Systematic literature review of usability capability/maturity models. Comput. Stand. Interfaces **55**(7), 95–105 (2018)
16. Ke, C., Xu, P., Wu, Y.: Maturity evaluation of emergency capability for earth-rock cofferdam construction. Hydropower Energy Sci. **35**(08), 94–97 (2017)
17. Tian, J., Zou, Q., Wang, Y.: Research on the maturity evaluation of government emergency management capability. J. Manage. Sci. **17**(11), 97–108 (2014)
18. Ji, H., Su, B., Li, M.: Study on the maturity model of emergency management capability in colleges and universities. China Work Safety Sci. Technol. **9**(01), 39–45 (2013)
19. Liu, Y., Zhao, J., Li, Z.: Study on the system and structure of earthquake emergency response capability maturity model. Earthq. Eng. Eng. Vib. **32**(06), 182–186 (2012)
20. Guo, X.: Research on the Construction of Evaluation Index System for Emergency Management Capability of Archives. Hebei University (2017)
21. Poghosyan, A., Manu, P., Mahamadu, A.-M.: A web-based design for occupational safety and health capability maturity indicator. Saf. Sci. **122**(9), 65–74 (2020)
22. Xie, S., Chen, Y., Dong, S.: Risk assessment of an oil depot using the improved multi-sensor fusion approach based on the cloud model and the belief Jensen-Shannon divergence. J. Loss Prev. Process Ind. **67**(11), 79–85 (2020)
23. Yan, F., Xu, K.: Methodology and case study of quantitative preliminary hazard analysis based on cloud model. J. Loss Prev. Process Ind. **60**(11), 116–124 (2019)

Author Index

Printed in the United States
by Baker & Taylor Publisher Services

Printed in the United States
by Baker & Taylor Publisher Services